Ghost Stories for Darwin

Ghost Stories
for Darwin

The Science of Variation
and the Politics of Diversity

BANU SUBRAMANIAM

UNIVERSITY OF ILLINOIS PRESS

Urbana, Chicago, and Springfield

Contents

The Red Queen Runneth

On Interdisciplinarity

"Well, in our country," said Alice, still panting a little,
"you'd generally get to somewhere else—if you run
very fast for a long time, as we've been doing."

"A slow sort of country!" said the Queen. "Now,
here, you see, it takes all the running you can do, to
keep in the same place."

—Lewis Carroll, *Through the Looking-Glass*

Traversing liminal spaces, traveling the hallways of academia, at the borderlands of disciplines . . . Almost there, but never quite. Meandering, half mesmerized, half muddled, always mumbling. Dare I speak? Almost there, but never quite. Almost a scientist, yet a feminist; almost a feminist, yet a scientist; almost a native, yet an alien; almost an alien, yet a native; almost an outsider, yet inside; almost an insider, yet outside . . . Almost there, but never quite. A life held captive in oppositions. How did I get to this tantalizing, much celebrated place, the home of the oxymoronic feminist scientist, this magical place of perpetual motion . . . nowhere, yet everywhere all at once?

This book has been many years in the making. The ideas have percolated, soaked, and marinated in the ethos of inter- and multidisciplines. It has been enriched by thoughtful and wise collaborators. It has been nourished by patient and generous colleagues. It has been ruminated over, meditated upon, and then reimagined and transformed multiple times. In many ways, this is a selfish project. This is the book I was looking for when introduced to women's studies while in graduate school in biology. While there is a lot of great work in feminist science studies, from the vantage point of an experimental biologist, nothing ever satisfied. I hoped someone would write it, but no one has. Believing that there was a genuine gap in the literature, I decided to continue this book project. It has been exhilarating and humbling to imagine a borderless academia while being housed and employed by a disciplined academy. I am well aware of the ways in which disciplines become necessary, sometimes in productive ways for the tools and scholarship they enable and at other times in unproductive ways when they limit or bind knowledge through institutional and intellectual structures.

And yet I have to remember the serendipitous paths that brought me here. It all began when I got on the plane at the Bombay Sahar International Airport on a warm, still August night, with visions of donning that revered white lab coat amid a sea of white-skinned, white-lab-coated male scientists studying the proverbial white male rat. The visions and dreams of white lab coats and white male scientists were so shaped by a postcolonial education that paraded the triumphs of dead white men that the incongruity of a large brown woman in a sea of white men scarcely occurred to me. Indeed, without the slightest sense of irony, I spent my early years working under a giant poster of Charles Darwin—my role model and idol.

I had crossed the oceans wanting to be a scientist. It had been comforting to imagine an intellectual life free of national, cultural, and social norms of gender, ethnicity, and sexuality. It seemed the ideal place for the tomboy, the third-worlder, the woman. Yet it was precisely the cultural and social norms of gender, ethnicity, sexuality, and nationality that came to be the most demanding factors in my pursuit of science. And so I knocked on the doors of women's studies in an attempt to understand my growing alienation.

For a while in graduate school, my training in evolutionary biology and women's studies remained exciting but separate and mutually exclusive. My doctoral work was on morning glory flower color variation. Even with my growing interest in women's studies in graduate school, the worlds of nature and culture remained totally separate. My graduate school life in biology was filled with morning glories and theories and experiments, but for studying culture and politics I had to travel, literally, from west campus (which housed the sciences) to east campus (which housed women's studies). While women's studies dealt with questions of biology, they were almost exclusively grounded in women's bodies or gendered representations of scientific objects.

This bifurcated existence continued for a few years until I found myself having coffee one day with a colleague in women's studies. The exact moment when I suddenly saw a bridge between my two worlds is one that remains vividly etched in my memory. A mentor (and now a treasured friend and colleague), Mary Wyer, asked me about my doctoral work. Usually colleagues in women's studies would respond with an "Oh, how interesting" and quickly change the subject. But Mary wanted to know more and I eagerly explained. I talked about morning glory flowers and the puzzle of their many varied flower colors. I explained my experiments to test whether variation in flower color was indeed maintained and which evolutionary mechanisms might explain their maintenance. She listened with interest and when I finished, she nodded her head with what seemed like perfect comprehension: "Oh," she said, "you work on diversity!"

This may seem obvious to some, but to me it was a momentous revelation! For the first time, someone had bridged for me the worlds of nature and culture, bringing together two parts of my academic life that had been until then discrete and separate. When I say "discrete and separate," I mean that in every way—geographically women's studies and biology were on two different campuses; intellectually they shared no common courses, texts, journals, or magazines; institutionally and socially there was no interaction between the two. My circle of friends in each were different, my experiences and memories never overlapping.[1] The very sights, sounds, smells, and textures of everyday life in each domain seemed unique, different, and distinct. To be sure, feminists did study science—but I never encountered a single one of them in my science department or work. And to be sure some of the scientists were feminists—but I never encountered a single one of them in my women's studies classes or research. To build a connection, I would have to do it myself. Now, suddenly, my colleague—by hinting that the ideas of variation in my doctoral work in biology was connected to terms such as *diversity* and *difference* in women's studies—helped me construct a bridge between the two. I knocked on the doors of women's studies looking for an alternative to the sciences. What I found was a rich set of theories that helped me understand my own alienation within the culture of science and reenergized me as a scientist. These theories were (and still are) powerful, providing a view of science as a set of historically derived practices and cultures that shape our study of nature. Thus, I began examining the genealogy of primary concepts in my subdiscipline, discipline, and, indeed, science itself. I began exploring how cultural practices and beliefs enter into scientific hypotheses, language, and theories despite the best intentions of individual scientists. I now realize that culture, politics, and discourse are inextricably intertwined in our understandings of nature, especially how scientific conceptions of gender and race profoundly shape the lives of those of us who are women, "foreign," and of color. Life has never been the same!

The irony of the situation always amazes me. For, rather than leave the sciences, women's studies offered me the tools to stay in science.[2] Feminist work allowed me to understand the processes that caused me to doubt my own abilities and dreams, and helped me develop a framework to recapture my earlier fascination with thinking and doing science. And I did indeed stay on in the sciences and finished my doctoral degree in biology, and in addition, a graduate certificate in women's studies. When I left graduate school, I left with two rich sets of training, credentials, and tools—one in evolutionary biology and a second in women's studies.

But how does one bring women's studies and evolutionary biology together? This question and challenge initiated for me a new research program in women's studies *and* biology, exploring how the feminist and the cultural studies of science could inform the experimental practices of science. Simply asking the question, giving it credibility as a question, tested some twenty years of focused training in science. Work in women's studies meant expanding considerably my disciplinary training, which since tenth grade lay exclusively in the sciences. But I had to ask: If we can have feminists with disciplinary training and scholarship in literary studies, anthropology, sociology, philosophy, history, and psychology, who all "do" women's studies, why not biologists? How can we build women's studies as a site with laboratories of its own? Working across the disciplinary divides of biological sciences and women's studies has meant confronting the differing objects of disciplinary study: nature and culture.

Nature, Culture, and Naturecultural Worlds

Defining the object of the biological sciences as nonhuman life creates the illusion of a human-free world, a world removed from ideology, politics, and culture. Conversely, human culture remains in the purview of the social sciences and humanities, a world removed from the natural. But what if we refuse this nature/culture binary? In coining the term *natureculture*, Donna Haraway challenges us to reject the binaries of nature and culture and attend to the constant traffic of discourses, information, and theories between the worlds of natures and cultures. There is no nature or culture, only natureculture.

Learning to explore the world from a disciplinary perspective is something to be learned, something practiced, something remembered. As Evelyn Fox Keller has argued, disciplinary training frames the kind of problematics a discipline is structured around (Keller 1996). In this book, I trace how I came to understand how disciplines engaged with the idea of variation through three sites: morning glory flower variation (tracing the genealogies of the idea of variation), alien plants and animals (tracing the geographies of variation), and the question of women in the sciences (tracing the biographies of variation).

In my training to be an evolutionary geneticist, a field of morning glories made me look and wonder: Why were there so many kinds of flowers? Why this variation? What maintains this variation? The dissertation that ensued examined whether, and how, natural selection worked to maintain flower color variation in morning glories.

Later, as a postdoctoral fellow and junior research scientist, the changing Southern Californian landscape made me wonder: Why the profound shifts in species compositions? Why were native species dwindling in numbers? The ensuing project tested ecological theories of natives/exotic plants and their soil communities.

Throughout this work, I have been haunted by the politics of diversity in the sciences, in particular, the question of women *in* the sciences. As a third world woman in the hallways of science, I was becoming increasingly aware of how bodies were read in scientific culture as having a gender, race, class, sexuality, and nation, and with these readings came assumptions about scientific potential. Women's studies introduced me to a long, well-documented, and persisting pattern of women's exclusion from the sciences. As I traced the history of flower color variation in morning glories, I wondered about the scientists. The key players were overwhelmingly white and from the west—the United States or Europe. Part I shows how their theories and theorization emerged from the social and political questions they were confronting in their geopolitical contexts. The biographies of scientists became a critical site in understanding the deep subjectivities of scientists and how biographies were connected to questions of genealogy and geography. I began examining my own identity and subjectivity as a postcolonial woman in the corridors of western science. These three case studies—morning glory flower color variation, invasion biology, and women in the sciences—animate this book and my exploration of feminist science and technology studies (FSTS).

Feminist frameworks are precisely what allowed me to recapture my earlier fascination for evolutionary biology. Once introduced to them, I found it impossible, undesirable, and unexciting to return to my idealized dreams of naive, apolitical scientists. Instead, my passion for scientific and feminist scholarship came together in ways I had never imagined possible. After all, as some feminists have pointed out, the world is not divided into objects and subjects. Rather, as Isabelle Stengers argues, the world itself is "open to interpretation" not because of us but because of the world's own activities (Stengers 2011). This dissolution of subject and object, of observer and observed, is one of the key insights of contemporary feminist science studies—taking naturecultural agency seriously is critical. I came into interdisciplinary work because I could not imagine doing science the "usual" disciplinary way. Science, as I imagine it, is not practiced in sterile hallways, removed from the social and political. I am constantly astonished at how we have vivisected a vibrant world into sterile disciplinary formations. Interdisciplinarity has taught me that there is another way to study the natural world—unbounded by disciplines but using the vast repertoire of disciplinary and interdisciplinary resources (not limited to the sciences). It is this vision grounded in the infinite and boundless possibilities of the feminist studies of science that has inspired and propelled my work. It is my hope, intention, and quest to entangle the natural sciences, humanities, and social sciences with one another, to see nature and culture as inseparable. In the feminism I want to practice, the sciences are a vital and critical part, a project where feminism and science, nature and culture, can sit at the same table trying to together understand and produce knowledge about the world.

Over the past decade, my work has sought to engage feminist science studies in the experimental practice of science. Bringing together the disciplinary methodologies, philosophies, practices, assumptions, methods, and languages of women's studies and the biological sciences has proven more difficult than I anticipated. Geographies of universities segregate disciplines, institutional structures make interdisciplinary careers difficult, and funding and publishing industries frown upon interdisciplinary work as unfocused and unrigorous. And perhaps the biggest challenge has been keeping up with the exploding research in multiple disciplines. Like Lewis Carroll's Red Queen, I have felt that I have to keep running to stay in the same place: perpetually "catching up." Working across two disciplines (women's studies and the biological sciences) that share virtually no spaces—geographic, institutional, or intellectual—has highlighted the myopia of disciplinary logics. I am struck by the compartmentalization of people and knowledge. At times, the only interdisciplinarity seems to be in my head! I find myself a walking incubator, terrified that I am nourishing the wrong kinds of ideas while starving the right ones. Telling right from wrong has meant a lot of running to stay in the same place.

This book is an attempt at interdisciplinarity in a very disciplinary world. It is purposefully a practical book—carving out an intellectual and experimental practice in the sciences that engages with the feminist studies of science. Its protagonist is clearly an experimental scientist recounting the process of coming to an interdisciplinary consciousness and practice, with all the joys and challenges that has entailed. It has been an exhilarating and humbling journey.

Acknowledgments

My curiosity about the biological world was nurtured by two middle school teachers, Mrs. Gowri Prasad and Mrs. Satyabhama. I have been enraptured ever since! My motto in life has been to surround myself with great and wonderful minds, in the hope that some of their ideas and passion will rub off on me! I feel particularly fortunate in having been enriched by so many patient and generous souls.

Thanks to the many colleagues who have read the entire manuscript and given me insightful and helpful feedback: Carole McCann, Rebecca Herzig, Michelle Murphy, Claudia Castaneda, Lisa McLaughlin, Sarah Richardson, Erika Milam, Jennifer Nash, and Kiran Asher. I am particularly indebted to Karen Cardozo and Mary Wyer for their careful and close reading and extensive suggestions for revisions. I have benefited from feedback from many who have read sections of the book: Angie Willey, Jim Bever, Peggy Schultz, Britt Rusert, Jennifer Hamilton, Ravina Aggarwal, Arlene Avakian, Anne Fausto Sterling, Karen Lederer, Rachel Lee, Laura Doyle, and Srirupa Roy. Sam Hariharan in particular has been that invaluable, gentle first reader in my moments of angst. I am indebted to Mike Dietrich and Laura Lovett for their advice and encouragement in developing the historical work in part I. The careful reading and helpful advice of two anonymous readers for the press considerably strengthened this manuscript. This book is much the richer for all of these wonderful people.

Many conversations with colleagues and friends over the years have helped shaped the arguments in the book: Angela Ginorio, Betsy Hartmann, Kavita Philip, Janet Jakobsen, Deboleena Roy, Evelynn Hammonds, Anne Fausto Sterling, Geeta Patel, Kath Weston, Charles Zerner, Donna Riley, Chikako Takeshita,

Virginia Eubanks, Miranda Joseph, Linda George, Priya Kurian, Laura Foster, Anjali Arondekar, Lynn Morgan, Rosemary Kalapurakal, Mike Whitmore, Devamonie Naidoo, Srirupa Roy, Itty Abraham, Ravina Aggarwal, Abha Sur, Jennifer Terry, Amir Alexander, Jane Lehr, Gwen D'Arcangelis, Claire Jens, Sangeeta Kamat, Rebecca Dunn, Vandana Date, Maralee Mayberry, and Lisa Weasel.

In navigating the intellectual and institutional challenges of interdisciplinary work, I am indebted to colleagues who have given me sage advice, an encouraging word, and cautionary notes at many key moments in my life. My gratitude to Mary Wyer, Sandra Harding, Angela Ginorio, Evelynn Hammonds, Anne Fausto Sterling, Karen Barad, Jean O'Barr, Afsaneh Najmabadi, Sue Rosser, Helen Longino, Evelyn Fox Keller, Cathy Middlecamp, Donna Haraway, Bonnie Spannier, Janet Jakobsen, Barbara Whitten, and Mary Sue Coleman. I am deeply appreciative of the many teachers who have helped shape my formative learning years, in particular during my undergraduate years, Drs. Meera Paul, Hannah John, Jilly Joseph, and Chitralekha Ramachandran, and in graduate school, Janis Antonovics, Mark Rausher, and Henry Wilbur.

Thanks to my colleagues at the UMass Amherst: Arlene Avakian, Laura Briggs, Alex Deschamps, Ann Ferguson, Tanisha Ford, Dayo Gore, Linda Hillenbrand, Miliann Kang, Karen Lederer, Nancy Patteson, Svati Shah, and Angie Willey; and at the University of Arizona: Miranda Joseph, Myra Dinnerstein, Janice Monk, Kari McBride, Julia Balen, and Sally Marston.

The University of Illinois Press has been a delight to work with. Many thanks to Larin McLaughlin, who has made this easier than I could have imagined; Dawn Durante and Jennifer Comeau for their help through the production process; and Maria denBoer for her timely copyediting.

As with most immigrants, networks of family and friends rather than roots or place are what have sustained me over the years. My parents, Radha Subramaniam and Shankar Subramaniam, have always encouraged my nonmedical, and nonengineering predilections, albeit with occasional bemusement and frustration! I could not ask for a more supportive parent than my mother, who reads most everything I write. My sister, Indu Ravi, has always been an inspiring and formidable act to follow. Thanks to the friendship of so many who have kept life interesting and adventurous: Sam Hariharan, J. Ravi, Karen Lederer, Kel Moorefield, Scout, Angie Willey, Karen Cardozo, Mary Wyer, Betsy Hartmann, Jennifer Hamilton, Arlene Avakian, Martha Ayres, Rebecca Dunn, Jim Bever, Peggy Schultz, and Kamala and V. Hariharan. While they are all a lively and supportive bunch, it was ultimately their annoying "You aren't done yet?" that propelled me out of my inclinations to procrastination! My deepest gratitude to all.

Interdisciplinary Hauntings

The Ghostly World of Naturecultures

> The more enlightened our houses are, the more their walls ooze ghosts.
>
> —Italo Calvino, *The Literature Machine*

> But I almost think we're all of us Ghosts . . . It's not only what we have inherited from our father and mother that "walks" in us. It's all sorts of dead ideas, and lifeless old beliefs, and so forth. They have no vitality, but they cling to us all the same, and we cannot shake them off. Whenever I take up a newspaper, I seem to see ghosts gliding between the lines. There must be ghosts all the country over, as thick as the sands of the sea. And then we are, one and all, so pitifully afraid of the light.
>
> —Henrik Ibsen, *Ghosts*

The woman in a flowing white sari glides by on a dark moonlit night surrounded by a fuzzy eerie glow. The trees rustle gently. A haunting melody plays in the background . . . These were frequent scenes in Bollywood and other Indian movies that I grew up with. Ghosts in Indian movies are always characters whose abbreviated life marks urgent unfinished business—usually characters that are murdered for the secrets they knew or for money, politics, or love. Not able to withstand the injustice of the murder, their ghosts appear at opportune moments in the film, often in song, to bring the past injustice to light. They always appear to those who, while frightened, are willing to listen, investigate, and excavate a gruesome and troubled past. The ghosts by their constant and insisting presence coax and shepherd the slow uncovering of the past, the uncomfortable but undeniable truth. In the end, the guilty are apprehended and the innocent consigned to eternal heterosexual bliss. The injustice corrected, the ghosts disappear into oblivion with peaceful and satisfied souls.

The realm of the preternatural in Indian culture always annoyed me. It flew in the face of life in postcolonial urban India suffused with an ethos of the secular, rational, and scientific. But in the movies, mythologies, comics, books, and popular culture the living and the dead, the natural, preternatural, and even the supernatural, mingled effortlessly. The film director and writer Ritwik Ghatak has insisted that the realm of spirits, ghost worlds, and reincarnations are quintessentially Indian; to understand India is to understand these worlds. But to my young nerdy science mind, ghosts signaled religious superstition and ignorance—factors that kept India a backward and third world country. Science and technology and the rational worlds they inhabited were the way forward to modernity.

These visions of a science removed from culture, politics, and history were quickly dashed once I was in the hallways of science. It was women's studies that helped me understand that, contrary to my earlier beliefs, scientific culture was not a "culture of no culture" (Traweek 1992). Rather, categories of gender, race, class, sexuality, and nation were everywhere, constantly shaping science, its practitioners, its cultures, and scientific knowledge. Over the past decade I have worked to translate this insight and my training in women's studies and the biological sciences into a combined practice—moving a disciplinary world of natures and cultures into an interdisciplinary one of naturecultures.

Studying naturecultures, however, poses a problem of methods. In traditional academic configurations, one studies the natural world by using methods of the natural sciences and the cultural world using methods from the humanities and social sciences. But how does one study a naturecultural world? How do you develop a practice where nature, science, *and* culture matter (Fausto-Sterling 2003)? The past three decades have yielded a wealth of discussion and debate about methodologies, ontologies, epistemologies, and methods in the feminist studies of science (Ahmed 2008, Alaimo and Hekman 2008, Barad, 2007, N. Davis 2009, Haraway 1988, 1997, Harding 1986, 1991, Hird 2004, Kirby 1997, Longino 1990, Roy 2004, 2008, Wilson 2004). One common theme spans these explorations—it is that knowledge is not out there to be discovered or is inherent in any scene or object, but that knowledge generation is an active engagement. Knowledge comes "to be" through complex processes including the scientists and their objects of studies, both located in their material, historical, geopolitical, economic, and naturecultural contexts (Stengers 2011, Barad 2007, Longino 1990).

For example, let us consider the famed and long-enduring debate on nature versus nurture. Recent scholars posit that this is a false binary. The phenotype, or the material body, emerges through the complex actions of nature *and* nurture. It is not as though nature acts through nurture or that nurture nurtures a nascent "natural" organism, rather that organisms are co-constituted and co-produced by nature *and* nurture, genes *and* their environments. As Evelyn Fox Keller maintains, it is not possible to apportion the percentage that we can ascribe to nature or nurture, although biologists and social scientists try mightily to do so (Keller 2010). Co-constitution and co-production suggest an inextricable interconnection (Reardon 2001). Reducing them into some original constituency misses the co-constituted knowledge of how matter comes "to be." We have to find a way to study them simultaneously. Indeed, the sooner we can give up the binary of nature and nurture and find new vocabularies to study the emergence and development of life, the better. A lofty goal, but a worthy project for feminist science and technology studies. Using the resources of the natural sciences, social sciences, and the humanities and arts, I narrate attempts at just such interdisciplinary naturecultural practices.

In attempting the bridgework, the more I tried to bring the worlds of women's studies and the sciences together, the farther they seemed to drift apart. Bruno Latour's formulation on disciplinarity is useful. In exploring the practices of modernity, he describes two sets of impulses (Latour 1993b). First, the "work of purification," whereby we create two entirely distinct ontological zones: that of the nonhuman/natural world on the one hand, and that of human/cultural world on the other. Alongside this work of purification, Latour also elaborates a second set of practices, namely, the "work of translation," which creates mixtures between the worlds/hybrids of nature and culture. Such is the paradox of the moderns, he argues; "the more we forbid ourselves to conceive hybrids, the more possible their interbreeding becomes" (Latour 1993b: 12). In retrospect, my task was more challenging because neither zone was all that stable. On the one hand, we carry on the work of purification in the academy by creating different sets of disciplines, tools, and methods and reify the separation of the worlds of human and nonhuman, nature and culture. On the other hand, this work of purification *creates* hybrids of all kinds between nature and culture. After all, while the staging of science might imagine pure scientists studying a pure nature, a naturecultural world is constantly making and remaking itself (Stengers 2011).

Feminist science and technology studies (FSTS) reveals a history of knowledge teeming with the interwoven worlds of science and politics, of biology and society, of nature and culture. Historians have argued that scientific knowledge should be understood within its historical, political contexts. After all, much of the scientific knowledge that we find questionable today (such as phrenology or craniometric studies about sex and racial differences or diagnosis of hysteria or neurasthenia, to just name a few) was widely accepted among the scientists of its times. This was "good" science done by "respected" scientists and accepted by their peers. As Nancy Stepan (1982) has argued in tracing the idea of "race," "the scientists who gave scientific racism its credibility and respectability were often first-rate scientists struggling to understand what appeared to them to be deeply puzzling problems of biology and human society. To dismiss their work as merely "pseudoscientific" would mean missing an opportunity to explore something important about the nature of scientific inquiry itself" (xvi).

As history reveals, scientific findings, and indeed all knowledge production, are emergent within the contexts of their production. Unless we are prepared to claim that contemporary science is inoculated against social and historical influences (surely a fanciful claim), it behooves us to be reflexive when it comes to our own knowledge production practices. Thus as science and technology studies has suggested, if science is "constructed," and we come to understand its workings, it can also be "reconstructed" (Woodhouse 2005, Campbell 2009). That is my goal in this book: to imagine a feminist reconstructive project for experimental biology.

What do I mean by "feminist"? Feminism today has expanded beyond a focus on women, to how material bodies, institutions, and structures are gendered. Feminists also recognize that woman is not a universal monolithic category, but always intertwined with other social categories such as gender, sexuality, race, ethnicity, class, and nation since women's experiences are shaped by their multiple social locations and identities. Thus, "feminist" means not just exploring women or gender, but recognizing the co-constituted meanings of categorizing human populations. We need to focus on the process of knowledge production, not just the content. Feminist historians of science also demonstrate that science has been central in the invention of biological categories such as sex, gender, race, class, nation, and sexuality, and subsequently reifying social prejudices in the inferiority of women, people of color, colonized, poor, or homosexuals. Historians remind us to examine the process of *how* this knowledge came to be. Such work requires critical reflexivity among knowledge producers. The goal of feminist science studies is not to produce knowledge that is prefigured in what it says about women or people of color or queer communities or third world nations (H. Rose 1997). *The goal is to develop* *an experimental practice and method that does not overdetermine or prefigure its conclusions.* The tensions and challenges of reflexivity and situating knowledge while simultaneously opening up new possibilities is the joyful acrobatics of academic play, one we must embrace with imagination and gusto (Stengers 2011). How do we develop such a reflexive and reconstructive project for experimental biology?

I am deeply influenced by Helen Longino's proposal that "we focus on science as practice rather than content, as process rather than product; hence, not on feminist science, but on doing science as a feminist" (Longino 1989: 47). Longino's rejection of a content-based approach comes from the dangers (surprisingly frequently in the literature) of conflating "feminist" with "feminine." In asking the question "Can there be a feminist science?" she argues that "if this means: is it in principle possible to do science as a feminist?, the answer must be: yes. If this means: can we in practice do science as a feminist?, the answer must be: not until we change the social and political context in which science is done" (Longino 1989: 56). I wholeheartedly agree. The challenge then is not a pointless battle between the irreconcilable frames of objectivity and social construction, but how to better understand the tensions between objectivity and belief as a necessary part of science and as central to the practice of science (Stengers 2000). In what follows in this book, I chronicle my attempt to craft a scientific practice that embraces the tensions between being a feminist (with all the commitments that entails) and being a feminist who recognizes the collusion between scientific knowledge and categories of difference. I do so by examining a central concept in both biology and women's studies—variation. I explore the histories and political contexts of the scientific production of the theories of variation and how they are deployed in three different sites: morning glory flower color variation, invasion biology, and women in the sciences. Since our focus must be not only on the resulting knowledge but also the practice—a contextual account of knowledge and its naturecultural co-production—this book chronicles the process of such work and its subsequently reconstructive imaginations. The binaries of nature and culture are deeply embedded in our academic institutions; in our disciplinary formations; and in our knowledge-gathering epistemologies, methodologies, and methods. To refuse this binary and produce knowledge about a naturecultural world is neither simple nor easy—the walls of disciplinary thinking, institutional barriers to publishing, finding jobs, and promotion and tenure are constant. Yet, I believe it is dangerous to suggest an easy reciprocity between nature and culture, between humans and nonhumans; our knowledge production has been far too mediated by the politics of the academy. While we can argue that science is in many places (Roosth and Schrader 2012)—in bacteria, plants, and animals—I am interested in exploring what is possible within the structures of academic science, and take seriously the experiences and agency of the practitioners of that science. Similarly, while science occurs in many sites such as in indigenous knowledge systems, or kitchen science, or DIY science, this book is an attempt to open up the epistemic authority and expand the practices and epistemologies of what we call "normal science." This book thus chronicles what is possible within the geographies of our academic institutions today while it imagines the fertile possibilities of a world not bounded by disciplinary logics.

The Ghostly Worlds of Naturecultures

As William Faulkner reminds us, "The past is never dead. It's not even past" (Faulkner 1951: 80). With this lesson firmly in mind, I began my forays into the interdisciplinary world of naturecultures. At first, I did not see the ghosts, not at all. Now I cannot believe that I ever missed them! Today they stand out in the landscapes around me. The many ghostly figures among the morning glories, the apparitions among the changing landscapes, and the luminous figures rendered invisible by the histories of science. I now see the hauntings, the ghostly apparitions, the shadowy silhouettes of our past appearing all around me, staring at me, insisting on being seen.

As I perched on the naturecultural bridge, the past and the present came clearly into focus and I began exploring. I began with tracing the history of the question of variation in evolutionary biology and suddenly new lines of connection came into view . . . And they appeared—the ghosts among the morning glories in North Carolina and the landscapes of Southern California and in the histories of women in the sciences. Tracing the history of flower color led me to the history of evolutionary biology, and the specter of eugenics emerged. In examining the landscape of Southern California and tracing the idea of the "native" in the history of ecology, the ghosts reappeared. And with the ghosts, returned the specter of eugenics. As I examined the history of women in the sciences, I encountered the *biopolitical scripts* (Takeshita 2011) where some bodies were deemed gifted with intellectual capabilities while others were relegated to kitchens, homes, wombs, menial tasks, or asylums. Again, *eugenic scripts* shaped biographies of those deemed science worthy and able "modest witnesses" (Haraway 1997). And the ghosts reappeared. In Bollywood tradition, these were the ghosts of the violently dispossessed, the unrealized talents, the forgotten brilliant, the sterilized, the mutilated, the unacknowledged, the ignored, the marginalized, the famous, and the common—whose lives were abbreviated by a brutal history, rendered invisible, whose genius was never realized, and whose voices were silenced by a disciplinary history devoid of people.

Who were these people, I wondered? What were their connections to the theories and methods I studied in my courses in biology? Why were they returning to haunt me? The Bollywood memories came back and I succumbed to the lure of the ghosts and the ghostly dispossessed. I began to scour the history books, to unearth the history I had never been taught. To my shock, I discovered that these were rather well-researched histories within the history of biology, just ones not taught to biologists. Similarly, women's studies did teach about the horrors of eugenics, but disciplinary silences never considered morning glories or the ecological landscapes as worthy objects of feminist inquiry. It was no wonder that these connections were not readily apparent—disciplines were constructed precisely to obscure their connections. With their mutually exclusive disciplinary objects firmly in place, the biological sciences and women's studies can go about their disciplinary pursuits undisturbed. Getting past the disciplinary silences, I dug in, and so emerged this book.

At the heart of all three of these sites—morning glory flower color variation, invasion biology, and women in the sciences—was the central question of variation. Related to variation are more popular terms: *diversity* and *difference*. How should we understand variation, diversity, and difference? Why is variation such a central question that it reverberates through the history of evolutionary biology *and* women's studies? Variable or not—what is the power in the idea? The central argument of the book is that disciplinary logics make invisible the political (eugenic) questions that are central to the question of variation. A naturecultural framework, however, makes visible why the question of variation is so central and long enduring in the cultural and natural worlds, and why the stakes are so high. Indeed, tracing the genealogy of the idea of variation reveals its fundamentally recursive structure—we return to the same question of variation again and again in the history of biology and politics. I have modeled this recursivity in the very structure of the book, as we revisit its persistence through examining the idea of variation through multiple sites in science—its genealogy, its geography, and its biography. The ghosts and their hauntings link these varied sites and show their naturecultural interconnections—it is indeed the same ghosts that haunt the interstices of our genealogies, geographies, and scientific biographies. Questions of genetic variation in human and nonhuman organisms are deeply linked to questions of diversity and difference in human populations steeped in tortured histories of slavery, colonialism, and genocide. *A naturecultural analysis reveals that the question of variation is fundamentally about power—the politics of life and death.* The seemingly innocuous history of genetic variation holds within it the countless bodies of the dead, the mutilated, the tortured, the irredeemable, the unwanted, as well as the brilliant talents that have gone unrecognized and unacknowledged—all those unknown, forgotten humans relegated to the rubbish heap of history. It is these wounded souls, these tortured histories, that roam the interstices of the question of variation. These are the ghosts of naturecultures . . .

Genealogies of Feminist Science and Technology Studies

Historians often date the origins of modern FSTS to Carolyn Merchant's 1980 book *The Death of Nature* (Schiebinger 2003). In a recent retrospective essay, I used "moored metamorphoses" as a metaphor to capture both the development of the field of FSTS as well as my own intellectual transformation (Subramaniam 2009). FSTS is a difficult field to define, as it is a heterogeneous and amorphous body of work that has emerged and grown organically rather than a field with consensus or cohesion. Its practitioners range from scholars in interdisciplinary programs such as women's studies, gender and sexuality studies, science studies, cultural studies, and visual studies to those located in traditional disciplines across academia. After early practitioners articulated and developed the theoretical foundations of a critique of the sciences, the field exploded. In many ways, I consider myself incredibly fortunate to have discovered the field when I did. When I knocked on the doors of women's studies, my mentors were able to introduce me to a vibrant body of work on feminism and science. The pioneers of the field had already established a foundational set of analyses, what came to be called the *feminist critiques of science.* The critiques made an important shift from questions of women *in* the sciences to articulating the relationship of gender *and* the sciences. The early works of Birke, Bleier, the Brighton Women and Science Group, Fausto-Sterling, Hammonds, Haraway, Harding, Hubbard, Keller, Longino, Lowe, Martin, Rose, Rosser, Schiebinger, Spanier, and Tuana (listed here in alphabetical order, not in order of importance) elaborated why women, gender, and feminism mattered in the production of scientific knowledge. The central critiques were directed at biological determinism, scientific objectivity, and assumptions about value neutrality, reproduction, and the labor of women; gendered images and language; challenging the boundaries between nature and culture; the role of capitalism; and the politics of knowledge and its production (Subramaniam 2009). From a core set of critiques, the field has exploded vertiginously across disciplines and interdisciplines, taking on a wide range of topics in almost every discipline with a vibrant and striking heterogeneity. In many ways this is the strength and promise of FSTS—there are no departments, journals, annual conferences, or organizing structures that bind or define the field. There is no established canon. Contributors have come from a wide range of places and from unconnected disciplinary and interdisciplinary origins and paths. In examining syllabi in the field, there appears no set of canonical readings that typify the field. While it is now difficult to keep up with the literature in the field, there is also something exhilarating and dynamic about the field's fluidity. Recognizing the dangers of unitary genealogies and citational practices (Hemmings 2011) and saluting this rich and heterogeneous history, I offer no single genealogy to the field.[1] Others have attempted to characterize it (Keller and Longino 1996, Lederman and Barsch 2001, Mayberry et al. 2001, Schiebinger 2003, 2008, Tuana 1989, Wyer et al. 2008).

Looking across these multiple genealogies, most recognize three sites or levels of analysis (Schiebinger 2003). First is the body of work on women *in* the sciences that chronicle the history, sociology, biographies, and activism of women scientists. Central to the field in its long and enduring history is confronting questions about whether variation in who practices science is good, and answering the question in the affirmative, the field has developed strategies to address women's underrepresentation. Second, scholars have organized around studies of the "cultures" of science (Knorr Cetina 1999, Latour and Woolgar 1986, Noble 1992, Traweek 1992, 1993). Women's participation in science and the contexts of knowledge production depend on the cultures scientists inhabit. Scholars have shown that science emerged as "a world without women" and that this history of male domination has fundamentally shaped the early cultures of science and lives on in contemporary scientific culture (Noble 1992). Science indeed presents itself as the ultimate privileged site immune to its unique history, geography, or genealogies. Scientific culture understands itself as a "culture of no culture" (Traweek 1992). Challenging such a view, feminists argue that histories of sex, gender, race, class, sexuality, and nation have deeply shaped and continue to shape the cultures of science. Finally, in addition to the practitioners of science and the cultures they work in, scholars have focused on the production of scientific knowledge. Here cultural critics have documented the myriad ways in which cultural understandings and mores of sex, gender, race, class, sexuality, and nation have historically shaped and continue to shape scientific knowledge. This work has motivated philosophers of science to examine scientific epistemology, systematically dismantling claims of objectivity and value neutrality, and, finally, proposing alternate epistemologies and methodologies for practicing science informed by feminist studies.

Feminist studies, like science and technology studies (STS), emerged from social movements committed to the political struggles of marginalized groups and activist struggles for a more just and inclusive world (Campbell 2009). Science, a powerful arbiter in the world, has been central to these histories of inequalities (and indeed also in their resistance). Despite the emergence of FSTS within women's studies, the field has yet to embrace the power of its own transformative critique. Women's studies remains firmly grounded in the humanities and the social sciences, and within women's studies, the sciences at best endure as a site of oppression and need for critique. This book builds on this rich legacy of FSTS. I approached this field as an experimental biologist interested in what feminism had to offer science. I soon discovered that *if science was constructed as a world without women, women's studies was constructed as a world without science.* Developing an interdisciplinary life across these fields has meant rethinking them both. A daunting task! This book is about this journey, engaging with the personal, the cultural, and the institutional, all critical to the project of knowledge construction.

In exploring the history of variation, the book argues that western societies (at least since Darwin) have struggled with biological questions of variation alongside political questions of diversity and difference. Nations and scientists aspiring to produce the "good society" engaged the sciences in that mission. But what is the "good society"—a diverse one with a multitude of people or one where only the best and the brightest reproduce to take us to new heights? These debates have haunted us for centuries. Science is one site in which these struggles with the interests of power are evident. These contentious processes, their histories and contexts, are often erased and scientific theories rendered "objective" with epistemic purity and claims to political and value neutrality. Delving into these histories reveals the profound debates around eugenics, about desirable and undesirable bodies—those doomed to sterilization, enslavement, or colonization or deemed perverse, deviant, pathological, or deficient. These histories, I argue, are evident in each of the three case studies, just under the surface. While we may debate selection pressures in evolutionary processes, or the eradication of invasive plants, or the wisdom of recruiting more women into the sciences, the spirit of the dreams of what we mean by a "good society" is at the heart of these debates.

In tracing the central and critical idea of variation in the history of ecology and evolutionary biology, I argue that the concept is one that connects biology and women's studies and the history of ecology and evolutionary biology. Comprehending these interconnections is central to understanding how the vocabulary of variation, diversity, and difference emerged and persists in both fields. I do this by exploring the idea of variation in three sites in three parts of the book—on genealogies, geographies, and biographies. I explore how variation is understood scientifically, is transformed geographically, and is embodied biographically. In part I, "Genealogies of Variation," I trace the idea of variation and the shifting understanding of its significance in the field of evolutionary biology. I explore how the idea emerged within particular political and economic contexts and has come to shape discourses on diversity and difference today. I excavate the centrality of eugenic scripts in our ideas of variation, and note the dizzying array of scientists who have participated in the eugenic project. I revisit my field experiments on flower color variation and reflexively examine how feminist studies of science can help us critique its goals, methods, and experiments. I suggest alternate ways in which morning glory flower color can and should be studied. I end with using fiction to imagine anew.

In part II, "Geographies of Variation," I use invasion biology as a site to examine how science theorizes plants and animal geographies, especially how nativism and nationalism shape ideas of belonging. It is no accident, I argue, that along with a panic about foreign plants and animals is a profound panic about foreign peoples as well, a panic that is distinctly gendered in reproductive fears of proliferation and miscegenation. I argue that this panic misplaces and displaces anxieties about globalization, labor shifts, and a fast-changing world onto a problem about the geographic origin of species. What is obscured by such panic are the economic and political interests that have ushered in large-scale environmental shifts through unregulated and large-scale overdevelopment. Drawing on four key moments in the past ten years, I develop a recent *biography* of the United States, a nation of interspecies migrants. I also examine collaborative experiments on the ecology of invasion biology. How easy is it to incorporate feminist, political, and activist concerns into experimental practice? I explore both the challenges and the possibilities. Finally, in a meditation on alien-ness, I examine my own standpoint as a foreigner in the United States studying foreign plants and animals and explore how these alien and kindred subjectivities may be a rich source of knowledge and meaning making.

Finally, part III, "Biographies of Variation," traces the case of women *in* the sciences and how ideas of variation and especially debates around them have shaped the histories of women participating in the sciences. The eugenic scripts that emerge in excavating a genealogy of variation reappear in recurring debates about women's intellectual and physical difference from men, a difference constructed against a western-white-male-as-norm backdrop. Rather than disrupting the gendered and racialized nature of science, programs for women in science largely focus on equity and parity in accommodating women in the structures of science rather than transforming science. The exclusive focus on increasing numbers of women has led to helping women achieve science's idea of a scientist rather than disrupting such normative ideals. Even transformative efforts such as female-friendly policies constructed around demands of pregnancy, motherhood, and family reinforce essentialized ideas of women, highlighting women as exceptional rather than transforming the normative ideals of science. Through biographies of scientists, I explore how FSTS can help move us beyond an equity feminist approach to women in the sciences.

From Variation to Difference to Diversity: Studying Naturecultural Worlds

Variation, diversity, difference—these are three central, critical, and foundational concepts that span women's studies *and* biology, key conceptual categories that permeate and bind the interstices of these interdisciplines. While the term *variation* sounds innocuous, scientific theories of variation come from contestations of very particular variation in humans—of sex, gender, race, class, sexuality, and nation—variation that was understood in a political syntax of its times. These histories have shaped and been shaped by science and are indeed constitutive of science. These were also the critical concepts and ideas that repeatedly emerged in my work on morning glories, invasion biology, and women in the sciences. At first I thought the overlap was fortuitous, but I now understand that such transdisciplinary convergences are never accidental. Instead, the idea of variation is deeply connected to varied sites.

In evolutionary biology, variation is the "stuff" or raw material of evolution. Starting with Darwin (and indeed before), naturalists recognized patterns of similarities and differences in organisms. Why such a wide variety of organisms, and why are some more similar to each other than others? For modern evolutionary biology, Darwin is key. Darwin, it is argued, moved us from earlier typological thinking where individual organisms were seen as varying ever so slightly from an ideal form, be they Plato's "ideal types" or Aristotle's essentialism (Mayr 1973, Sober 1980). Rejecting a metaphysical ideal type, Darwin emphasized the ontology of concrete things (individuals) and the ways in which individuals varied ever so slightly from other individuals (Ghiselin 2005). This move from metaphysics of "typologies" to an empirical science of variation or "populations" is a key conceptual and epistemological shift of Darwin (Mayr 1973, Sober 1980, Grene 1990). Scientists developed empirical categories and methods to measure variation. While Darwin himself was deeply steeped in ideas of gender, race, and nation of his time (Gould 1980, Hubbard 1990), Darwinism opened evolutionary biology and the development of theories of evolution to the powerful forces of social Darwinism and the politics of gender, race, class, nation, and sexuality of nineteenth-century Britain. Science—its theories, methods, and knowledge—emerged as a critical site where politics and ideologies of difference were encrypted into the very foundations of the field. As the field emerged, physical, behavioral, and intellectual differences between groups were biologized—from differences in skeletons, skulls, metabolic systems, and reproductive organs to traits such as intelligence, feeblemindedness, pauperism, and alcoholism (Allen 1996). At the heart of this ideological project rested the logic of eugenics. If those born "good in stock, hereditarily endowed with noble qualities," could be encouraged to breed, as Francis Galton, Darwin's cousin and the father of eugenics, suggested, the human race could be improved. Concurrently, preventing the "bad stock" from breeding could hasten improvement (Allen 1996). The purpose of eugenics, Galton wrote, "is to express the science of improving stock, which is by no means confined to questions of judicious mating, but which, especially in the case of man, takes cognizance of all influences that tend in however remote a degree to give the more suitable races or strains of blood a better chance of prevailing over the less suitable than they otherwise would have had" (qtd. in Allen 1996: 23).

To analyze the large volumes of data that emerged, new mathematical and statistical techniques were needed. Thus the field of biometry emerged. Biometrics was "generated as a toolbox for eugenics, as a collection of measurements of characteristics and correlations for the analysis of inheritance" (Louçã 2008: 655). Biometry was not a field that emerged separately and was then applied to the eugenic project; rather, it emerged to expressly aid the eugenics project. The foundations of eugenic thinking, its methodology, and its theories and statistical modeling are thus deeply indebted to the innovations of biometry and its "scientific form of social Darwinism" (Norton 1983). Eugenics was also a culture, providing a bridge between different fields, exerting a powerful influence for defining an agenda for biology and the social sciences where they would inform each other. "Statistics was the language used for that communication and it was a new language—society and not only individuals mattered" (Desrosières 1998: 68, Louçã 2008).

With the motivations of a eugenic project, biologists ventured into large experiments on measuring various characteristics and developed the statistical techniques to evaluate the patterns of variation. Not surprisingly, differences between various groups emerged. Historians have documented the ways in which some populations emerged from these efforts as superior to others (Anderson 2006, 2008, Birke 1999, Bleier 1984, Fausto-Sterling 1985, 1995, Hammonds 1999, Hammonds and Herzig 2009, Haraway 1989, 1991, Harding 1993, Hubbard 1983, 1990, Markowitz 2001, Philip 2004, Prakash 1999, Raina and

Habbib 2004, Reardon 2005, S. Richardson 2012, Sayres 1982, Schiebinger 1989, 1993, Stepan 1985, Terry 1999, Verran 2001). Thus new ontologies and the "logics" of superior sexes, races, nations, and sexualities emerged, and these drew their authority from the growing power of "science." It is critical to recognize that these ideas of difference were "biological" differences and believed to be innate, shaping the character and biological potential of individuals within populations. These were claimed to be essential characters of groups, not constructed or easily mutable. Indeed, these logics took hold in the logics of colonialism and in the political cultures of colonial powers as well as members of the colonies. Thus seemingly innocuous ideas of variation were converted to profound and political ideas of difference between various groups at various times in history—men versus women, whites versus blacks, colonial powers versus the colonies, heterosexuals versus homosexuals, elite versus working poor (Birke 1999, Bleier 1984, Briggs 2002, Hammonds 1999, Hammonds and Herzig 2009, Haraway 1989, 1991, Harding 1993, Hubbard 1983, 1990, Markowitz 2001, Philip 2004, Prakash 1999, Raina and Habbib 2004, Reardon 2005, Rubin 2012, Sayres 1982, Schiebinger 1989, 1993, 2004, Stepan 1985, Terry 1999). *The benign language of variation is thus converted into the profoundly political language of difference.* Indeed, difference now stood for the superiority of some groups and the inferiorities, pathologies, deviance, and perversions of others. The differential biological potential of different groups and arguments about biological difference (rather than social inequality) emerge as the main cause for social problems of the superiority and inferiority of various groups. Reading eugenics through a history of sex, gender, race, class, sexuality, ability, and nation reveals the profoundly political project that it was. In uncovering these deeply political histories, and tracing questions of identities and institutions, I discovered a profound logic of difference. This logic is articulated through policies on individuals as well as institutions; the logic of difference also disciplines institutions to act in particular ways (Foucault 1977). Darwin moved us from one kind of essentialist thinking in typologies, but the shift to variation and population thinking ushered in a new essentialism of identity—of sex, race, class, sexuality, nation, and ability—that lives on in contemporary politics (Gannett 2001, Grene 1990). Essentialism has a long legacy of being dangerous for the "other," and locating human capacities and capabilities in the immutable body was a profound development (Said 1979).

Drawing on the work of historians and philosophers of biology, feminist critics of science have argued that the institution of science reified social power and privilege into claims of natural and biological difference. As identity-based movements reclaimed the power of the margins—of femininity, blackness, queerness, disability, and third world nations—a negative politics of difference was translated into a positive politics of diversity. Thus, diversity emerged as an aesthetic of celebrating cultural variation and expressions of cultural difference within liberal discourse (Eriksen 2006). New social movements dedicated to multicultural and inclusive visions converted difference into a celebration of diversity. The problem, they point out, is not that groups are different from each other. Rather, the problem is in judging and evaluating groups whereby some group differences are translated into the superiority of one group over other groups (Lorde 1984).

Many scholars have argued, however, that contemporary work on diversity, rather than questioning differences, often "mirror" and reinforce them (Baez 2004). Indeed, as Chandra Mohanty maintains, in practice, diversity has become a discourse of "benign variation, which bypasses power as well as history to suggest a harmonious empty pluralism" (Mohanty 2003: 193). Diversity in its recent institutional incarnation has been utterly domesticated and depoliticized and largely seen as "good" because it has lost its political roots of structural issues of sexism and racism. Diversity is used in diverse ways to mean very different things (Ahmed 2012), reminding us that categories are contextual and unstable, and deeply political. As Omi and Winant claim, identity categories like "race" do not have content by themselves. These categories, emerging at particular historical moments, were mobilized toward clear political and economic ends. Race is a "formation" through the complex and shifting sets of political projects that organize human bodies and social structures in the aid of particular agendas (Omi and Winant 1994). The challenge then is to explore how and why our discourses of difference organize the category of difference within particular configurations of power and how they shape individual experiences and attitudes (Baez 2004). How were ideas of variation translated into the violence and virulent politics of the inferiority of difference in sexes, races, classes, nations, sexualities, and abilities? Through the development and use of Darwinian natural selection, variation was translated into difference, and into the conceptual and political landscape of sex, gender, race, class, nation, sexuality, and ability so familiar to us today. Population thinking retains and refigures inequalities in new "scientific" ways, obscuring their political genealogies (Gannett 2001, Grene 1990).

Intersections of Interdisciplinarity

In examining feminist histories of identity categories, two factors are striking. First, interdisciplinary explorations across biology and women's studies reveal profound intersections in the production of difference and identity. Intersectionality is most often defined as "the relationships among multiple dimensions and modalities of social relationships and subject formations" (McCall 2005: 1771). While I am aware of the contested histories and critiques of the idea of intersectionality, its multiple meanings and heterogeneous uses, its peculiar institutionalization, its intellectual and political mobilization, and the fears that its depoliticization and domestication in the academy have foreclosed its more radical potential (Crenshaw 1991, Hammonds 1994, Hancock and Yuval-Davis 2011, McCall 2005, Puar 2007, 2012, Nash 2008, 2009, Yuval-Davis 2006), my ambition is more particular and rather modest.

I use the term *intersectionality* to highlight the creation and naming of particular material bodies—such as the "heterosexual black slave woman" or the "Oriental woman" or the "white homosexual man"—produced through the histories and theories of science. Indeed, understanding intersectionality as a process emerging through the institution of science and the material production of different bodies has not been adequately explored or theorized (Collins 1999).

Second, the histories of sex, race, sexuality, class, nation, ability, and various categories of identity are not mutually exclusive and did not arise independent of each other. By this I mean that while new social movements have highlighted universal identity categories such as woman, black, colonized, LGBTQI, queer, disabled, or third world (and thus created their normative counterparts in man, white, straight, able-bodied, and first world), historians of science suggest otherwise. Rather than emerging as universal categories, identity categories appeared through material bodies in very particular and deliberate ways. The politics of biological differences created hierarchies of difference—of nuanced and intersecting biological potentialities of various material bodies.

While we often view categories such as sex and race as empirically and biologically distinct, they are murky. Perhaps the best researched and most written about is the case of the South African black woman Sarah Bartmann, or the "Hottentot Venus," who was paraded across Europe as a scientific curiosity and whose body was probed and catalogued during her life and beyond. Her open sexuality and unfeminine behavior provoked anxiety among white western scientists, who ultimately pathologized this body, which destabilized white male superiority at home and abroad (Adams and Pigg 2005, Stoler et al. 2008, Takeshita 2011). Much has been written about this case, and the analyses powerfully and systematically present how the colonial scientific authority systematically "produced" and created the inferior colonized "savage" black African woman's body (Fausto-Sterling 1995, Gilman 1985, Magubane 2012,

Sharpley-Whiting 1996, Takeshita 2011). The extraordinary scrutinizing of Sarah Bartmann that endlessly chronicles the intense racism and sexism that accompanies her life also redeploys racism and sexism in its retelling, cautioning us that revisiting history can never be innocent (Fausto-Sterling 1995). Nancy Stepan (1986) has pointed out that the historical record shows repeated and direct analogies made between identity categories. For example, she demonstrates how the race-gender analogy was used in the nineteenth century to claim women's smaller brains and protruding jaws as evidence of their evolutionary inferiority to European men. She describes a process where "by analogy with the so-called lower races, women, the sexually deviate, the criminal, the urban poor, and the insane were in one way or another constructed as a biological 'race apart' whose differences from the white male, and likeness to each other 'explained' their different and lower position in the social hierarchy" (Stepan 1986: 40–41). By the mid-nineteenth century, racial biology became "a science of boundaries between groups and the degenerations that threatened when those boundaries were transgressed" (Stepan 1985: 98). In a similar vein in "Pelvic Politics," Sally Markowitz argues that the very idea of sex differences emerged from racialized meanings of bodies. She traces how the category and ideology of sex/gender rests "not on a simple binary opposition between male and female but rather on a scale of racially coded degrees of sex/gender difference" (Markowitz 2001: 391). Rather than a universal or species-wide man or woman, evolutionary biology produced the "manly European man" and the "feminine European woman" as evolutionary types that mark the pinnacle of a hierarchical project of race (Markowitz 2001: 391). The very categories of identity emerge as co-constituted and co-dependent intersectional categories. Thus, characteristics like femininity are racialized themselves in the historical record; "femininity" could not be assumed to be a characteristic of *all* females (Schiebinger 1993). The distinct categories of sex and race cannot be presumed to be historically evident. Cathy Gere (1999) demonstrates this in a paleontological study where scientists confronted fossils with two distinct sizes of skeletons. Do these represent a sexually dimorphic population with differently sized males and females or two differently sized populations, one large and another smaller? She argues that only in making assumptions of racial and sexual differences can such a history be interpreted. Assumptions about race and sex are deeply intertwined.

What emerges from such different ontologies of race and sex is a profound political "logic of difference" that comes to underlie the foundations of medicine and science in creating racialized and gendered bodies (Hammonds 1999). For example, in tracing the origins of gynecology, historians have shown how racialized understandings of black women as "strong" (as opposed to elite white women's fragility) translated into an extraordinary biological capacity to bear pain. Thus for the father of gynecology, Marion Sims, "blackness defined what pain meant for these women, just as white womanhood defined his understanding of white women's pain experience" (Wanzo 2009: 159). Such logic of difference permeates nineteenth-century racial taxonomies and racial differences in creating what has been called the "medical plantation" (Dudley 2012). It is from such logics that black women's bodies were subjected to vaginal surgeries without anesthesia (Wanzo 2009). These logics presented unique paradoxes. Black women's difference in their high threshold of pain provided the logic for surgery without anesthesia while at the same time their sameness assumed knowledge about their bodies to be transferable to white women. What is striking and irrefutable but unsurprising about this history is that the logic of difference always benefits the elite and powerful at the expense of the poor and marginalized.

The histories of eugenics and gynecology amply demonstrate the profound consequences of the logics of difference—in the untold pain and suffering for those deemed inferior and/or expendable. The political nature of the scientific knowledge becomes particularly clear when we trace how such knowledges were put to use. For example, in the late nineteenth century, educators such as Kenneth Clarke warned against the higher education of elite white women (Sayres 1982). Drawing on biological theories that women's biology necessitated energy for reproduction, they argued that women's intellectual development would necessarily come at the cost of reproduction. Drawing on racist and xenophobic arguments, Clarke argued that if elite white women were given access to higher education, their low fertility would lead to a nation overrun with immigrants. Women's access to higher education was thus opposed on biological grounds to save the elite white race (Sayres 1982). If indeed such biological theories were deeply believed, we see no evidence of mass campaigns for the education of immigrant women, poor women, or women of color whose purported reproductive activity yielded such race panic!

The power of the logic of difference underlying scientific claims of the varying capacities of human bodies lies in the politics of biological determinism. Scientific claims of biological differences were not understood to be mutable but rather fixed and determined by biology. The foundational work in FSTS has documented how scientists "produced" biological differences of gender, race, class, and nation—claims of anatomical difference, physiological differences, as well as differences in behavior, temperament, and intellect. These biological differences are naturalized by culture. Feminists have amply demonstrated the tautological arguments and the evolving circular logic as knowledge travels between the natural and cultural worlds. Fausto-Sterling (1987) sums it up well in the title of one of her essays: "Society Writes Biology/Biology Constructs Gender." Biology and society, nature and culture, are inextricably intertwined as culture constructs ideas of gender and biology naturalizes those societal ideas and mores. Cultural and social norms are now rendered scientifically objective and biologically immutable.

Given these histories, understanding intersectionality as a material-semiotic process through which bodies—endowed with sex, gender, race, class, sexuality, nation, and ability—are produced is critical. The metaphor that best describes this history is the game of *Jenga*. Dismantle any one of these blocks/categories, and the whole edifice comes crumbling down. Feminist and antiracist scientists have valiantly challenged simplistic notions of sex and race (and we will see more of this history throughout this book). Scientific theories are literally constituted through the material politics of privilege and marginalization, through the practices of determining who reproduces and who is sterilized, of who is welcomed into a nation or kept out of its borders, of who counts as an "individual" in the eyes of the nation honored with the right to vote, property, industry, or representation. Scientific politics help constitute social politics. It is at these intersections of biology and politics that the ghosts of naturecultures come tumbling out.

THE QUESTION OF GENRE

Narrative is at the heart of this book—narratives that the sciences and humanities have come to tell about the world. One of the striking patterns that emerged in telling the story of variation is its recursive structure—we encounter our struggles with the question of variation again and again. This pattern of recursion structures the book in the ghosts that haunt my narrative again and again. This is rather fitting as science was once a branch of philosophy and great scientists such as Galileo and Newton thought of themselves as "natural philosophers." Given that the term *scientist* was coined as recently as 1834 by William Whewell, it is rather extraordinary how deeply and stubbornly the idea of science as divorced from the humanities has taken root (Pigliucci 2012).

One of the challenges in writing across the fields of women's studies and the biological sciences involves the question of genre. How best to narrate the idea of variation and diversity than in the very form of the book? Feminists have long been troubled by the notion that disciplinary genres of writing are necessarily rigorous. Rather, codified genres within disciplines are a product of their individual histories. Taking this to heart, I employ multiple genres to examine the theoretical quandaries in the field in crafting a book rich in its form and function. This book embodies the varied methods and methodologies multiple disciplines offer, an interdisciplinary text that literally sits between experimental humanities and experimental sciences. I draw from the vast repertoire of method and methodologies across the natural and social sciences, humanities and arts. In addition to traditional experimental scientific methods and historical and rhetorical analyses, I also weave in autobiography, auto-ethnography, and fiction.

The personal narrative, the use of the "I," is another site of discomfort and suspicion. In the sciences, it marks subjectivity and a lack of rigor. Yet, the historical record is replete with examples (as this book amply demonstrates) where purported objective research in the sciences that assiduously cultivates a passive voice is in fact profoundly subjective and biased. In the humanities and women's studies in particular that ushered in the importance of the personal as a site of theory making, the "I" is now viewed as overused and the personal story as dangerously standing in *as* theory (Walters 1996). Yet, the personal narrative I have found is one that eases interdisciplinary conversations and best articulates what is at stake for knowledge production as well as the biographies of individual researchers. Since one of the goals of the book is to nurture new biographies, new possibilities of professional identities and disciplinary formations, I have embraced the personal narrative as an imaginative site that helps us connect questions of women in the sciences, the culture of science, and the knowledge that individual scientists produce.

Finally, I have taken to heart recent calls for a nonhuman-centric approach to STS (Haraway 2008, Kirksey and Helmreich 2010, Margulis 1998, Margulis and Sagan 2002). To this effect, I approach the nonhuman actors not as passive objects in experiments whose true natures are "revealed" through experiments, but as living agents who shape the stories that can be told. Indeed, they help co-construct scientific knowledge, although their participation—why and how they can speak, and if they can be heard—is shaped deeply by the methods used. I have worked hard to engage the nonhuman world and to listen to and learn from the multispecies voices that make up our world.

Tracking Ghosts: Hauntings from a Eugenic Past

I have to laugh. Years after this work, having traversed the breadth of academe with interdisciplinary meanderings that have taken me through the richly diverse hallways of knowledge that academia offers, after a robust scientific training, I now see hauntings. As Salman Rushdie succinctly summarizes, "Now I know what a ghost is. Unfinished business, that's what" (Rushdie 1988: 540). I spent most of my childhood railing against the superstitious invocation of ghosts. They filled Indian folklore, mythologies, and the rich tapestry of Indian cinema. The past could never be left alone, never forgotten, the earthly life never partitioned from the netherworlds. And here, in the hallways of rational science, in the beauty of flowers, in the changing landscapes of Southern California, in the all too scientific worlds of science and scientists, I have taught myself to see ghosts. Gory, bloody, violent ghosts from our eugenic past.

"What kind of case is a case of a ghost?" asks Avery Gordon in *Ghostly Matters*.

It is not a case of dead or missing persons *sui generis*, but of the ghost as a social figure. It is often a case of inarticulate experiences . . . of more than one story at a time . . . In the case of modernity's violence and wounds, and a case of the haunting reminder of the complex social relations in which we live . . . what can represent systematic injury and the remarkable lives made in the wake of the making of our social world. (Gordon 1997: 24–25)

How do you learn to see ghosts? And what do you see once you learn to see them? This has been my project in FSTS. Trained in evolutionary biology, I saw a field of morning glories and asked about flower color variation. I did not ask why it was the most obvious question. The landscapes in Southern California provoked me to ask questions about native and foreign species, without questioning the blurry distinctions between the native and the alien and the histories of the plants. The problem of women in the sciences elicited strategies to increasing their numbers, without any questioning of the gendered and racialized expectations of science.

Years later, I look at the same fields and see the ghostly apparitions of a eugenic past—the many mutilated, tortured, imperiled, and dead bodies, the stigmatized, contained, disciplined bodies of communities and nations of color, the poor, those deemed mentally incompetent, inferior, the many lives deemed not worth living. In tracing the genealogy of variation, all these histories came tumbling out.

Eugenics today evokes the holocaust, racial hygiene, genocide, mass sterilizations of peoples considered "inferior," the horrors of the unholy alliance of science and politics. Eugenics was without doubt an important and fundamental aspect of many key movements in the past two centuries, intimately linked to ideologies of race, nation, and sex, and also a part of several institutions such as population control, social hygiene, state hospitals, colonial governance, and the welfare state (Dikötter 1998). Yet, eugenics, it turns out, has had very different biological and political valances at different periods, embraced by an astonishing number and range of scientists with diverse political persuasions. Eugenics was thus less about a clear set of scientific principles than a "modern" way to discuss social problems in scientific terms; politicians and scientists of various political persuasions appropriated eugenic discourse to promote social policy under the guise of objective and apolitical language of science and the laws of nature (Dikötter 1998). Indeed, this book is an attempt at moving from an idea of "natural selection" to a "naturecultural selection"—in understanding both the naturecultural contexts of our organic evolution as well as the evolution and politics of scientific theories and their histories.

A genealogical exploration of the idea of variation in the field of ecology and evolutionary genetics lays bare the history of eugenics, social Darwinism, and neo-Malthusianism. The question of variation in early evolutionary biology was closely intertwined with questions of undesirable, unwanted human variation—the idiots, imbeciles, paupers, feeble minded, deformed, promiscuous, epileptics, criminals, and alcoholics. The rest we know well. The history of forced institutionalization, of sterilization, population control, and the horrendous eugenic practices across much of the world. Since World War II most of evolutionary biology has clearly and unequivocally distanced itself from eugenics. But rhetorical flourishes with little historical exploration or reflection can never exorcize all the ghosts. Indeed, eugenic ideas live on. We have also seen a resurgence of biological claims of sex, gender, race, class, sexuality, nation, and ability (Fisher 2011, Paul and Spencer 1995, Bliss 2012, Roberts 2012). The ghosts live on in almost all aspects of current biological practice. Learning to see them is not just about seeing the ghosts, seeing the history, the political and cultural legacy of the field, but about laying bare the epistemological and methodological apparatuses that have framed our seeing for more than a century.

Remember, invisible things are not necessarily "not-there." Learning to write about the invisible is "about how to write about permission and prohibitions, presence and absence, about apparitions and hysterical blindness. To write stories concerning exclusions and invisibilities is to write ghost stories. To write ghost stories implies that ghosts are real, that is to say, that they produce material effects" (Gordon 1997: 18). In learning to see ghosts, scientific practice transforms into a deep-seated historical practice, where the objects and subjects of science and their histories come hurtling into focus. But it is this kind of seeing that disciplinary thinking has systematically excluded.

Drawing from Zora Neale Hurston, Avery Gordon writes: "Ghosts hate new things . . . The reason why is because ghosts are characteristically attached to the events, things, and places that produced them in the first place; by nature they are haunting reminders of lingering trouble. Ghosts hate new things precisely because once the conditions that call them up and keep them alive have been removed, their reason for being and their power to haunt are severely restricted" (Gordon 1997: xix). Indeed, evolutionary biology has stayed within the same eugenic scripts of biological determinism and essentialist thinking, reproducing the hierarchies of sex, gender, race, class, sexuality, and nation. Reading the history of science is like playing whack-a-mole. Each time a claim of biological determinism has been dismantled, another one rises up. Each time, FSTS has pointed to the poor sample sizes, faulty logic, reductive methodologies, poor methods, or unwarranted conclusions. Yet our newspapers and scientific journals are filled with renewed claims about sex differences, racial differences, sexuality differences, and national differences. The politics of sex, gender, race, class, sexuality, and nation continue to haunt the construction of scientific knowledge.

I return to my naive conceptions of the beliefs of an apolitical and value-free institution when I first entered the hallways of science. The "culture of no culture" turns out to be entirely about culture, identity, and difference, as does scientific knowledge. What has possibly been the most humbling discovery is my recognition of the symbolic power of the preternatural. Ghosts, rather than a superstitious legacy of a past, are a haunting reminder of an ignored past. Rendering ghosts visible and learning to listen to them attentively is a lesson about the unacknowledged and unresolved injustices of history. Living with ghosts forces you to confront the past, or the dead never go away, history never sleeps, the truth can never be erased, forgotten, or foreclosed by modernity. As long as we stay embedded in the scripts of disciplinary thinking that have haunted evolutionary biology for the past two hundred years, the ghosts will continue to haunt us.

PART I

Genealogies of Variation
The Case of Morning Glory Flowers

It is interesting to contemplate an entangled bank, clothed with many plants of many kinds, with birds singing on the bushes, with various insects flitting about, and with worms crawling through the damp earth, and to reflect that these elaborately constructed forms, so different from each other, and dependent on each other in so complex a manner, have all been produced by laws acting around us . . . There is grandeur in this view of life, with its several powers, having been originally breathed by the Creator into a few forms or into one; and that, whilst this planet has gone cycling on according to the fixed law of gravity, from so simple a beginning endless forms most beautiful and most wonderful have been, and are being, evolved.

—Charles Darwin, *The Origin of Species*

CHAPTER ONE

Thigmatropic Tales

On the Politics and Social Lives of Morning Glories

weeds without value
humorous beautiful weeds.

—Mary Oliver, *Morning Glory*

A morning glory at my window satisfies
me more than the metaphysics of books.

—Walt Whitman, *Leaves of Grass*

An academy with separate and distinct disciplines has carved knowledge production into unique objects of studies and methodologies, obscuring the teeming life between the worlds of natures and cultures. A central goal of this book is to illustrate what an interdisciplinary *naturecultural* analysis would look like. In this chapter, I revisit my doctoral work on the maintenance of flower color variation in morning glories to explore how a feminist analysis can help explain the shape and scope of my research. How did it come to take the particular shape and form that it did? Might a naturecultural approach to flower color variation be different? Indeed, one of my claims here is that an interdisciplinary education would have fundamentally reshaped my work on the evolutionary biology of morning glory flower color variation. Inspired by the touch-sensitive thigmatropic tendrils of morning glories, which allow the plants to scale large objects and burrow into narrow crevices, I narrate tales of morning glories through the curious and adventurous tendrils of naturecultural storytelling.

Many a North Carolina Summer Morning . . .

During fieldwork, I most remember the spectacular sight of morning glories on a warm summer morning. Splashes of purple, white, blues, and pinks in intense, dark, and light hues, all entangled. The flowers were fresh, open, shimmering in the morning dew. The bees and insects buzzed around them. By noon, the flowers withered under the blistering sun. Perhaps the bees had done their magic or perhaps the plants had fertilized themselves. By afternoon there was virtually no color in the field. By evening it was entirely a field of green. The following morning, another glorious vision. Despite the intense heat of southern summers, and even after years of backbreaking and sometimes futile fieldwork, the morning vision of morning glories always could stop my heart. They do indeed make the morning quite glorious!

Fifteen years later, I look at the same field of morning glories and see the ghostly apparitions of a eugenic past—the many mutilated, tortured, imperiled, and dead bodies, the stigmatized bodies of communities and nations of color, the poor, those deemed mentally incompetent, inferior, the many lives deemed not worth living. I see the methods, theories, and statistical models developed in the aid of eugenic goals, and the patriarchs who helped shape the discipline of evolutionary biology and hence my own disciplining and enculturation in the world of science: Darwin, Galton, Malthus, Pearson, Fisher, Dobzhansky, Haldane, Muller. Most had strong connections to ideas we now recognize as eugenic, connections that have been slowly forgotten or judged to be irrelevant.[1] In tracing the genealogy of variation as an idea in the history of evolutionary biology, all these histories came tumbling out, the ghosts appearing in all their historical fineries. And feminists peeked out here and there in this history, intertwined in social policies that had complex and surprising twists and turns. This, then, is a story of how I came to examine the history of evolutionary biology, my horror and wonder at what my biological education had revealed and concealed, and my growing maturation into the complexities and contradictions of history and its implications for experimental practice.

 Learning to see ghosts (and now haunted by my increasing inability *not* to see them) was itself an intellectual education. It meant understanding *why* I saw what I saw and why, after fifteen years of a different kind of education, I have come to see the morning glory fields so differently. I see these ghosts— of feminism, science, and politics—as revelations of what disciplinary structures conceal. Interdisciplinary work reveals what disciplinary histories have rendered invisible. Many years later, after meandering through the richness of intellectual pathways in the academy, I realize that disciplinary work—epistemology and ontology—is learned behavior.

An evolutionary biology of flower color variation allows us to ask questions about flowers in a field. It ignores how we came to see the field and the question of flower color variation as *the* obvious question. In evolutionary biology today, histories of race, gender, and sexuality are rendered invisible, indeed irrelevant to the question of flower color variation. Instead, as I began theorizing the field of morning glories and their variation as a graduate student, I began with the requisite scientific apparatus of clear hypotheses, making visible logical argument, good experimental design, appropriate statistical analyses, and careful interpretation of results. The ontology of flower recognition—the categories I came to recognize such as the color of the flower, its hues, its petal shape, and its reproductive anatomies and predilections—are things I *learned* to see as significant. Training to be an evolutionary biologist involved a "disciplining" to identify the "right" problems, learning to ask the "right" questions—being enculturated into seeing and observing in the particular ways of the field. The ontological categories such as flower color are "right" because they are seen as critical to evolution and the forces of natural selection. With this training, I came to look at a field of morning glories and wonder: Why all these flower colors? Why are they all not purple? What keeps the white flowers reproducing year after year, but always just around 10 to 15 percent? Why do you never see an entire field of white flowers? I cannot say that these questions were new. Many had asked them before, and I read their explorations with interest and these questions engaged me too.[2]

Naturecultural Worlds

My work on morning glory flower color variation was grounded in the study of one of the most fundamental concepts in evolutionary biology—variation. In women's studies and ethnic studies, we pay a great deal of attention to the related concepts of diversity and difference. Did theories about plant diversity have anything to do with theories of human diversity? My training in evolutionary biology featured a history with scant attention to the humans that shaped the field; the focus was instead on the theories, experiments, models, and statistical techniques that were all in the service of purportedly explaining the nonhuman natural world. Inspired by feminist science studies and with a naturecultural view in mind, I began to explore the history of evolutionary biology and morning glories. This approach sparked a new set of questions: Why and how does one come to study a research object such as morning glories in a field of evolutionary biology, where human histories are largely deemed irrelevant? What if we consider such histories relevant? What can they teach us? Revising morning glory flower color variation as a naturecultural problem, I quickly discovered that the idea of variation in evolutionary biology has been intricately linked throughout history with ideas of human diversity and difference. In this chapter I revisit my doctoral research experiments testing evolutionary theories of flower color variation. In tandem with rethinking the experiments, I began to explore the biographies of "famous founders" in the field of evolutionary biology, becoming familiar with the historical debates in which modern theories on variation emerged. In the following chapter, I expand this analysis to trace the genealogy of the idea of variation through the history of evolutionary biology. In tracing this history, we will see that it takes researchers considerable work to understand the materiality of variation (as factors, genes, alleles, and DNA sequences), and that the history of the idea of variation is one that fundamentally travels from the worlds of nature (plants, animals, human biology, and subsequently molecular worlds) to the worlds of culture (ideas of diversity, difference, and inequality) and then back to the worlds of nature and then culture in endless loops. Indeed, the idea of the naturecultural would suggest that these are artificial distinctions and each is constitutive of the others. But telling this history necessitates that we hold on to the binaries of the natural and cultural worlds; I demonstrate how I came to discover these binaries and then unlearn them in a fragmented and disciplined academy. As I delved deeper into the historiography of eugenics, I discovered multiple histories. Internalist histories, largely told within evolutionary biology, ignore the political motivations of scientists or the social impact. Critical histories focus less on the science and more on the motivations of scientists and the social impact. Reading these contrasting histories, I came to again experience the ghostly presence of eugenics within the sanitized narratives of the disciplines.

The Genetics of Morning Glory Flower Color

Morning glory, *Ipomoea purpurea*, is an annual summer vine that grows about 2 m in length. It belongs to the family *Convolvulaceae*. It has heart-shaped, alternating leaves and, as its name suggests, its most distinctive feature are its flowers. Shaped like a funnel, the flower consists of five fused petals. The flowers are primarily pollinated by insects, especially bumble bees (Defelice 2001), but they are self-compatible and so can pollinate themselves.

A little digression here about morning glory flower color. How scientists go about understanding flower color variation brings us to the question of epistemology (or how we go about "knowing" what we know) in evolutionary biology. For contemporary evolutionary biology, the critical answers to flower color variation lie in following the genetics of morning glory flower color variation because evolution selects a phenotype (outward appearance) of flower color and its corresponding genotype (particular variants of genes or alleles).[3] In insect-pollinated species such as morning glories, the visual cues of the flower are critical. So, to understand the patterns of phenotypes of flower color that are visible in the field, we need to understand the underlying genotypic variation to the phenotypes. Understanding how natural selection works at each of the individual flower color loci will help us see how natural selection maintains the variation of flower color that we can observe. If one is interested in evolution, flower color variation is an ideal model system in which to explore the actions of evolution (Clegg and Durbin 2000).

The genetics of flower color, it turns out, is not so simple. There appear to be multiple loci (the region of a chromosome that holds genes for a particular character is called a locus; plural loci) that determine the final color of a flower. Furthermore, at each locus, most populations have multiple variants of the gene (called alleles) circulating, creating floral polymorphisms. At least four main loci are believed to determine flower color variation.

- The p locus determines whether a flower is purple or pink. There are two main alleles at this locus: P and p (blue is "dominant" over pink).[4] Therefore PP yields blue flowers, Pp yields blue flowers, and the recessive pp yields pink flowers.
- The w locus determines the intensity of the pigmentation and here the two alleles, W and w, are co-dominant—WW (dark color), Ww (light color), ww (white color). The white-colored flowers have rays of blue or pink depending on the alleles at the P locus.

Two other loci work very differently:

- The i locus works epistatically with (i.e., in concert with) the W locus to create a brilliant blue or pink. If a plant is ii and WW, then the flower will show intense blue or pink pigmentation with a velvety sheen.
- The a locus works epistatically on all other loci. If a plant is a recessive homozygote, aa, then it works epistatically to shut off all the other loci, so the flower is albino—pure white with no rays at all.

Contemporary evolutionary biology, like much of the biological sciences, is a reductionist science by design—grounded in the belief that breaking down a complex phenomenon into its individual parts and understanding each of the parts will yield a complete understanding of the whole. Understanding the world of natural selection at each locus, this approach argues, helps us understand what is happening to the whole flower and its color. Evolutionary biology foregrounds narratives where the visible traits we see—phenotype or frequencies—are the product of the complex workings of natural selection at the genotypes of the different loci that determine that particular trait. Therefore, to understand the splashes of color in an open field, one needs to understand how natural selection is working on these various attributes of color—color of the flower, its hue, iridescence, and so on. Most likely, the stories at each of the four loci—p, w, i, a—are different. A good experiment would be designed to understand the workings of natural selection at each locus separately. For the purposes of this chapter I focus on the w locus.

THE EXPERIMENTS

Here, I discuss two experiments that animate this work (Subramaniam 1994). Both deal with the w locus. The first focuses on the frequency of the two alleles W and w and the second on the fitness or relative performance of the three genotypes at the W locus—WW, Ww, and ww.

Frequency of W/w

The w locus determines the intensity of flower pigmentation. Two alleles circulate in most populations, W and w, and they are co-dominant, producing dark-colored flowers with darker rays (WW), light-colored flowers with dark rays (Ww), and white flowers with dark rays (ww). The color of the flower (and rays) depends on the alleles on the p locus. W and w are said to be co-dominant since both alleles express themselves in the heterozygous state (Ww). Surveys of fields show that w ranges in frequency from zero to 15 percent and it is almost never higher than 15 percent. Why not? What limits the w from increasing in frequency and what forces work to keep w as an extant allele—why is it not eliminated? Why do we not have W as the only allele in morning glory populations?

   I began with four field sites. All the experimental sites were selected so that there were no morning glory fields in the immediate vicinity. This was to ensure that no pollen (or only miniscule amounts) from other the experimental plants would affect the experiment. Two experimental treatments were created. Two of the fields were planted with high frequencies of the W allele and the other two with high frequencies of the w allele. The seeds for the experiments were generated through an elaborate series of genetic crosses so that all individuals—WW, Ww, ww—came from a similar genetic background. Therefore, any differences we saw could be attributed to the w locus. Plants on the four field sites were allowed to grow, flower, and go to seed over the growing season until the frost killed the plants. All the seeds were collected, and then the frequency of the two alleles W and w in the next generation (i.e., the seeds) were ascertained by growing them in the greenhouse. And then the moment of suspense: seeing whether the fields would increase, decrease, or maintain their initial frequencies of alleles. And the results were exciting!

We discovered that there had been a shift in the frequencies of the alleles (Subramaniam and Rausher 2000). Populations that began with high frequencies of the W allele showed an increase in the alternate allele w, and vice versa. In the experiment and across all four sites, the minority allele for flower color intensity increased in frequency, explaining our observations in natural populations where flower color variation continues to be maintained, although the exact ratio varies by location. Here was experimental evidence for a case of "balancing selection" acting on the w locus, so the action of natural selection at high frequencies of either locus gave the minority allele a selective advantage, therefore "balancing" the variation of alleles. Which flower color did better depended on its frequency in the population, that is, it was frequency dependent. Selection favored the less frequent. It explains why neither allele is ever fixed or eliminated. One explanation is that bumblebees, the plant's primary pollinators, favor dark colors. In contrast, the white flowers that get fewer bee visits are more likely to self-pollinate (Shu-Mei Chang and Rausher 1999). Self-pollination of white flowers gives the w allele a selfing advantage (R. A. Fisher 1941), that is, alleles of a plant that self-pollinates have a frequency advantage in the next generation. The pollinator preferences therefore may play a key role in maintaining variation at the w locus. Results such as these are not that common. A clear demonstration of changing allele frequencies is not easily obtained. It was exciting to see such a clear shift in allele frequencies.

Fitness of W/w

The w locus produces stunning differences in dark-, light-, and white-colored flowers (pink or blue). Fields of varied hues of pink and blue are quite breathtaking. Previous experiments had suggested that there may be a heterozygote advantage, or heterosis, at this locus, that is, the heterozygote Ww has higher fitness (produces more seeds or larger seeds) than the two homozygotes WW or ww. Whenever a heterozygote has higher fitness, the argument goes, both alleles W and w are preserved. In addition to balancing selection, this is a second evolutionary mechanism by which variation can be maintained. In this experiment, the three genotypes were produced through a series of crosses where the genetic background was randomized. Seedlings of the three genotypes were planted and then allowed to grow throughout the season and the seeds were collected from the plants. The seeds were then grown in the greenhouse and their genotypes ascertained through test crosses. What the experiment showed was that there was indeed heterozygote advantage (heterosis), that is, plants that were Ww produced seeds that were significantly larger than those that were produced by either WW or ww plants. Since seed size and weight correlate positively with the size and fertility of the plants they produce, this experiment shows that Ww has a selective advantage. Of course, once Ww is selected for, both alleles W and w are selected—again suggesting the actions of balancing selection through heterosis and another potential mechanism through which variation at the w locus is maintained (Subramaniam 1994).

THE EUGENICS CONNECTION

The above experiments show two main results. First, the distribution of flower color was not the result of random chance. Rather, natural selection worked to actively maintain genetic variation in flower color. Second, the maintenance of variation was enabled through the actions of natural selection, in particular, balancing selection (through the mechanisms of frequency-dependent selection and heterosis at the w locus). The very frame of the experiment—Why are there so many colored flowers? What is maintaining the variation? Is it there just by random chance? Or is selection actively working to maintain this variation?—comes from a longer history and debate about the nature of variation and selection in evolutionary biology. Steeped in evolutionary biology and its theories, the visions of morning glory flowers do evoke particular theoretical frames. Fully understanding the results of the experiments and their significance required me to explore a very particular debate in the history of evolutionary biology, a debate fundamentally grounded in the history of eugenics.

THE CLASSICAL/BALANCE DEBATE

Why is variation such a critical problem in evolutionary biology? From a naturecultural view, the debate about the nature of variation and selection that animated my experiments comes from an important debate in evolutionary biology. I had indeed studied these as central questions in my evolutionary biology classes. What I had not realized and what was absent in my textbooks and scientific papers were the political questions, in particular, eugenic questions that animated the scientific debate. There are two key scientists who are relevant here, both significant figures in evolutionary biology: Hermann

J. Muller and Theodosius Dobzhansky. Both of them were committed social activists and scientists and devoted to using scientific knowledge to better human societies. "Although a biologist may do his research on mice, *Drosophila* flies, plants or bacteria," Dobzhansky argued, "the ultimate aim should be to contribute toward the understanding of man and his place in the universe" (qtd. in Paul 1987: 334). Muller similarly wanted to use science, believing that "men should eventually be able to control the process, even in themselves, so as to greatly improve upon their own natures" (qtd. in Paul 1987: 321). Despite their shared passion for social betterment, they fundamentally disagreed on what kind of ideal society to develop, reflecting the complex history of eugenics. Eugenics was a broad philosophy about human betterment and was embraced by a wide range of political ideologies, including the political left and the right. Muller wanted a world that would be filled with brilliant men such as Newton, Lenin, Pasteur, Beethoven, Omar Khayyam, Pushkin, Sun Yat-sen, and Marx (Muller 1984). His many proposals for improving society included starting an organization that collected sperm of great men and made them available to women (Paul 1987). A homogeneous world of brilliance was the pinnacle of a great society for Muller. In contrast, Dobzhansky held that diversity was the "supreme value"; he favored a world replete with social, political, and genetic variation. "Do we really want to live in a world with millions of Einteins, Pasteurs and Lenins?" he asked and answered with a "No!" (Dobzhansky 1962: 330). To him genetic diversity and cultural diversity were related—just as he wanted a world that was culturally diverse, he believed that it was good for individuals to be genetically diverse, or heterozygous, and for populations to be polymorphic (Paul 1987).

These political and ideological positions spilled into their science, as for both Muller and Dobzhanksy, scientific and social values were inextricably interconnected. Muller's ideas were shaped by World War II, the use of the atomic bomb, and the effects of radiations seen in the horrors of postwar Japan. Muller was a pioneer in the research demonstrating the harmful effects of radiation. To him radiation produced mutations and variation that were inherently undesirable. Dobzhanksy, in contrast, saw variation as desirable and linked to a deep commitment to a diverse and egalitarian world. Along with L. C. Dunn, he published *Heredity, Race and Society* (Dobzhanksy and Dunn 1952), which used scientific evidence to counter biological claims about race. While they acknowledged that "not all human beings are alike," they cautioned against pure biological or cultural determinism (Gromley 2009).

To Muller, all mutation was deleterious and thus an active eugenics program was necessary to foster the purifying effects of natural selection. Mutations, according to Muller, were harmful, and accumulating mutations increased what he termed the "genetic load" of a population. With the backdrop of post–World War II Japan, Muller promoted the view that we should reduce mutations and variation. In contrast, Dobzhansky saw variation as generally adaptive and thus the presence of variation was a sign that natural selection actively maintained genetic variation through balancing selection. Because he felt that populations with the highest genetic variation would have an advantage over those that did not, he believed that the more coercive practices of eugenics would not work. To Dobzhanksy, genetic variation was critically important for evolution and genetic diversity was something we should cherish. Dobzhansky favored a world replete with diversity (genetic, cultural, and political), while Muller strove for a world where we were all smart and kind (Paul 1987). Again, scientific and social values are intimately linked.

This disagreement on whether variation was good or bad fueled a major controversy in the 1950s and 1960s in evolutionary genetics on the genetic variability of natural populations, the nature of selection, and the genetic effects of atomic radiation. This was nicknamed the classical/balance debate. Would the increase in mutations through radiation increase the genetic load of populations? The relationship between genetic variation and natural selection was the key issue. Muller and Dobzhansky figured prominently in the debates, with Muller taking the classical position and Dobzhansky the balance position (Beatty 1987, Dietrich 2006).

Dobzhansky and Muller were both interested in the amount of genetic variation available for evolution to act on and the impact of radiation-induced genetic variation. The question revolved around the amount of over-dominance, or heterozygosity, in populations (Crow 1987). In the classical position promoted by Muller, changes lead to the gradual substitution and eventual fixation of more favorable alleles, genes, and chromosome structures over the less favorable. Therefore, most loci according to the classical position should be homozygous and heterozygotes should be rare. A number of factors could cause this—mutations may be neutral, deleterious mutations are eventually eliminated by selection, or perhaps the mutations are selected for their beneficial effects and eventually fixed (Dobzhansky 1955). In contrast, the balance position promoted by Dobzhansky held that most loci should be heterozygous. Homozygotes would still occur but would not be as prevalent because Dobzhansky believed that heterozygotes had an inherent selective advantage. Dobzhanksy cast himself as primary advocate of balance position (Dietrich 2006). So which was true? Answers to this question revolved around the relative number and importance of heterozygotes (Dietrich 2006). Despite efforts to bring it to a close, the classical/balance controversy stalemated. The key questions surrounding variation, or difference were: Do we want a nondiverse, homogeneous world filled with a few brilliant types, or do we want a diverse heterogeneous world filled with cultural, biological, and social variants? What do we do with those in the fringes of normality or social acceptability? These questions still haunt biology as these competing visions of a uniform versus diverse society continue to be debated, as we will see in the following chapter.

The introduction of molecular biology only altered the classical balance controversy rather than settled it (Dietrich 2006). Some of the foundational questions of whether variation is good or bad, whether it is nature or nurture, continue to haunt biology. A genealogy of variation in the following chapter will demonstrate that these foundational questions have endured two hundred years of biology even as knowledge builds and changes. The same questions of homogeneity and heterogeneity, of sameness and difference, of nature and nurture, persist. My experiments on morning glories thus follow in the footsteps of this history of classical/balance debate.

GHOSTLY PRESENCES

So what if the founders of evolutionary biology were eugenicists? Does this mean that biology is forever connected to its roots and doomed to play out its eugenic scripts? Yes, I believe it does, *unless* biologists deal with this history rather than choosing to ignore it entirely. I have come to realize that while I innocently gazed at a field of morning glories, this history shaped what I saw and the questions I asked. Why this variation? Why so many colored flowers? While my biological education revealed particular answers, I now realize why and how I framed them in this way. It is worthwhile to ask: Why am I asking these questions and not others? Why have these two perspectives remained so compelling? The binary scripts of good and bad variation are tied to a long history of a binary politics of eugenics where scientists wrestled with the ideal vision of a good society—should it be one with homogeneous brilliance or heterogeneous diversity? It is these two positions that shaped my vision of the morning glory field, my seeing of flower color variation as a problem, and consequently my designing experiments to ask whether variation was maintained or eroded by the actions of natural selection.

One of the striking aspects of this history is how overtly political and ideological some of the founders were. The inability of evolutionary biology to resolve this debate may be a consequence of mixing politics with science, but I think not. While some scientists continue to be engaged in using scientific knowledge toward political ends, the norm is to reassert claims to political neutrality. While the sciences go through periods where politics are visible and periods when it is never overtly discussed (Paul and Spencer 1995), I would argue that by not recognizing that the originary formulation of the field was so grounded in human differences we are doomed to revisit the same questions unless or until we engage these groundings. While evolutionary biology after the horrors of the holocaust was more careful about the use of "race" in the post–World War II era, it is clear that marking "race" as a category endures today. For instance, in recent years the FDA approved its first "ethnic" drug (Kahn 2005, Fullwiley 2007, D. Roberts 2008). As Lisa Gannett reminds us, ideas about biological "race" have been recoded in the terminology of population thinking (2001). The ideological commitments remain but are expressed in a new vocabulary. While this history is seldom addressed in the training of evolutionary biologists, the history and foundational political debates are nonetheless deeply embedded in the theory and methods of biology. It is this history we need to unearth, revisit, confront, and silence.

A Social History of Morning Glories

Reflecting on my morning glory experiments, and in the context of a budding awareness of the cultural ideologies embedded in debates within my field, I explored another set of questions. These questions followed from a sense that scientists were all too willing to describe phenomena in terms of mutually exclusive categories: good/bad, natural/unnatural, human/nonhuman. Who were the actors in my morning glory experiments? What of the plant itself? What of its agency? Its own history? How have its social history, its migration patterns, and its entanglements with humans shaped and continue to shape its biological evolution? Asking such questions about morning glories, taking seriously the notion that nature and culture are not neat categories, provoked me to cast my experimental object in a new light.

Morning glory is an ideal naturecultural plant, a "companion species" (Haraway 2007) whose evolutionary and migratory history is deeply implicated in human histories. *Ipomoea purpurea* belongs to the genus *Ipomoea* and the family Convolvulaceae. Largely considered a pest and weed in agricultural contexts, it nonetheless has its fans and a vibrant social history. The word *convolvulus* comes from the word *entwine*, reflecting the vine-like morphology of many of its species. The word *Ipomoea* has similar origins from Greek, where *ips* means "a worm" and *omoios* means "resembling," describing a worm-like plant with a twining habit (Defelice 2001). The specific name *purpurea* refers to the most common color of its flowers. In literature morning glories are a symbol of farewell and departure and also a metaphor for momentary brilliance—someone or something thing that starts impressively but fades quickly.

The annual vine is believed to have its origins in the highlands of Central Mexico (Barkley 1986, Mabberley 1997). Domesticated by pre-Columbian people, possibly in association with maize and bean culture, the remarkably diverse flower is believed to have been selected for its aesthetic appeal. Collected from the Aztecs and sent back to Spain by Hernán Cortés, the vines were planted in monastery gardens (Defelice 2001). Subsequently, sixteenth-century European explorers sent samples from the new world to Europe for classification and study. *Ipomoea,* the largest genus of family Convolvulaceae, has at least five hundred species. The genus includes *Ipomoea batata,* the sweet potato, which was to become a famous food plant, as well as *Ipomoea purpurea,* the tall morning glory that has become a much loved and favored ornamental vine in Southern European gardens and Great Britain. It has since adapted to tropical, subtropical, and warm temperate regions of the world. Growing in the wild, this self-propagating plant is found across the globe in North America, Central America, South America, Australia, India, Africa, and tropical and sub-tropical regions of the world (Auld and Medd 1987, Srivastava 1983). While it is not clear exactly when it was introduced to the United States, the historical record shows that it has been cultivated since at least 1629, and morning glories became popular in gardens of eastern North America by the early eighteenth century. By this time, the plant had also escaped cultivation to become a weed. Today it is a weedy plant species along roadsides and disturbed habitats as well as in fields alongside corn, soybean, and tobacco (Defelice 2001). It can be found across the eastern and southern United States, all the way west to California, south through central and South America, and as far north as Ontario and Nova Scotia. Throughout most of the world, it is tended as an ornamental flowering plant, growing in gardens, in hedges, and on fences (B. D. Morley and Toelken 1983).

Its admirers ensured its wide and global distribution as human migration patterns doubtless shaped its evolutionary history. Some argue that it is surprising that the genus is so well and equally represented across most continents on the Planet since it is believed to be a more recently derived genus (A. Mc-Donald 1991). Others suggest that it may be one of the earliest global genera. Two hundred and fifty million years ago the world was fused into one large landmass, Pangaea. After this supercontinent broke up, splitting Eurasia from the Americas, each of the proto-continents had vastly different configurations of plants and animals. While insects and birds were more easily able to migrate across the two land masses, a few plant species did as well, one a morning glory relative, the sweet potato, *Ipomoea batatus* (Mann 2011). Morning glories can also be prolific. One tall morning glory can produce 26,000 seeds (Crowley and Buchanan 1982)! The combination of producing seeds prolifically and its intimate association with human migration means it is widely distributed across the globe, makes morning glories a wonderful naturecultural object of study with rich global and local histories (A. McDonald 1991, Glover et al. 1996). For example, studies demonstrate that southeastern populations of morning glory show reduced genetic diversity in biochemical and molecular polymorphisms for flower pigments, most likely due to genetic bottlenecks in the southeastern U.S. populations compared to those in Mexico (Glover et al. 1996).

Morning glories are known for their striking diversity in flower color. There are as many as twenty-one distinct phenotypes of flower color that have been documented. This stunning diversity is likely due to horticultural escapes in the eighteenth and nineteenth centuries (Glover et al. 1996). The plant is simultaneously beloved and reviled. Most books on gardening have a section devoted to growing morning glories and also a separate section on how to control them (Defelice 2001).

Morning glory has emerged as a common term to describe a range of events. Its beauty and simplicity evokes spiritual and ritual celebration (Moi 2009). It is a syndrome associated with retinal detachment and optical anomaly (De Laey et al. 1985) as well as a wind pattern. It also appears in books, paintings, photographs, poetry, songs, and soccer (Brabazon 1998) as well as serving as a metaphor for individual lives and biographies (Gordon Chang 2001, Dahl 2001). It has been used in folk medicine for stomach ailments and as a laxative. It has a cult following because of its famed hallucinogenic properties, although it can also be toxic. The Internet sports many websites with guides and growing instructions for various species of morning glories. But along with its ornamental beauty, it is most written about for its weedy character across the world. Morning glories are competitive with economically important plants and crops (Buchanan and Burns 1971, Crowley and Buchanan 1982, Oliver et al. 1976). Through several centuries of growing alongside cultivated crops, and facing ever sophisticated methods of weed control and eradication (Elmore et al. 1990), the morning glory has also faced evolutionary selective pressures (Chaney and Baucom 2012).

And indeed these naturecultural travels and histories of *Ipomoea purpurea* shaped their agency in the experiments I conducted. The morning glories in North Carolina have histories with the local agricultural crops, in particular corn and tobacco. I remember driving field to field talking to local tobacco farmers trying to collect morning glory samples that I could use in my experiments. I wanted to get as many "wild" samples as possible and went seeking them. In every case I got astonished looks and responses, including a question whether I had come from the local lunatic asylum! Why on earth would anyone want this dreaded weed? They of course encouraged me to take every last one from their fields. The emergence of tobacco in North Carolina is connected to the Civil War. It is said that northern soldiers raided local tobacco barns and developed a taste for the local tobacco, creating a wide demand for North Carolina tobacco after the war (Duke Homestead Tour 1990). This led to a flourishing industry whose wealth helped fund the very institution I was studying morning glories in.

Is this social history a side story? Perhaps cute, maybe interesting, but really nothing to do with morning glories? Or perhaps while details of its social history are illuminating, they are already captured in the experiments? At this point, I had to ask myself—why did I choose to study morning glories as an evolutionary biologist? The answer surely is to understand the biology of morning glory populations as a model system that sheds light on evolutionary forces and their shaping of the natural world. And here, I would argue, social histories are important. A plant that has been in the United States since the early 1600s must have a rich evolutionary history. When and how did the species, the source of the experimental plants, arrive in North Carolina? How are these histories similar across the southeastern United States? Or is there something peculiar to the local histories of tobacco and other crops in North Carolina that have shaped the histories of the particular plants I used in the study? In the scientific literature, results are often presented as species-specific rather than contextualized in local histories. Considering local contexts and histories and the complex naturecultural evolutions of morning glories surely would provide a more precise and complete account of evolution? Yet such questions are always considered outside the scope of contemporary biological research.

As knowledge about morning glories has grown, they have become the choice model in many laboratories, emerging as an ideal plant model system in ecological genetics (Baucom et al. 2011). What is also illustrative about the sociology of science is how model systems come to be constructed. Flower color is a wonderful trait through which to understand the workings of evolution—especially ones where the genetic mechanisms are better understood. Others have explored the opportunities as well as the challenges of model systems (Valent 1990, Rader 2004, Jansson and Douglas 2007). While model systems enable a deeper and more precise understanding of one system, they are also pernicious. As particular and peculiar varieties get created, the particular genotypes that are selected for the study are often shared globally for experiments. The emerging model system increasingly produces knowledge about organisms that is utterly removed from their local ecologies and evolutions. While we may eventually produce copious amounts of data of organisms in the laboratory, they often tell us little about the original intent of the biological sciences—understanding nature and its complex evolutions.

IN THE THEATER OF SCIENCE

In discussing the classic experiments of Louis Pasteur, Bruno Latour (1993a: 140) reminds us that experiments are theatrical events—constructed very particularly by the scientists so as to let the object "speak" and "reveal" its truth.

> What is an experiment? It is an action performed by a scientist so that the non-human will be made to appear on its own. It is a special form of constructivism. Who is acting in this experiment? Pasteur and his yeast. More exactly, Pasteur acts so the yeast acts alone. We understand why it is difficult for Pasteur to choose between a constructivist epistemology and a realist one; he creates a scene in which he does not have to create anything. He develops gestures, glasswares, protocols, so that the entity once shifted out, becomes automatic and autonomous.

If the experiment were constructed in a different way, different facts and truths emerge. Experiments need to be understood as "constructed" and we need to recognize that *both* scientist *and* the object of science (morning glories in this chapter) are important players in this theater. What understanding emerges about morning glories does so through the particular intra-actions of scientists and experimental subjects (Barad 2007)—they together produce the "truth" about morning glories. To the extent that science fails to attend to this, science uncovers only a fraction or at times only a fiction of the story.

The results of my morning glory experiments were significant, showing that patterns of flower color were not random but rather the acts of balancing selection. But the results were produced through a carefully constructed and contrived experiment. The experiments were designed to be "good" experiments where the results were definitive with clear and easy interpretation, and where ambiguity was minimized. In this sense, the experiments were well designed. But what exactly do these experiments reveal? In order to be positive that the effects are *solely* due to the variables in question, experiments need to be carefully constructed. This involves artificially producing the experimental plants with known genotypes as well as carrying out the experiments in controlled and sometimes artificial contexts. In these particular experiments, in order to produce the experimental plants that were genetically identical (except for precise variation in the w locus), several steps had to be undertaken. For instance, genetic crosses were carried out in plants grown in the greenhouse away from the uncontrollable lives of pollinators; seedlings were first grown in the greenhouse and then transplanted in the field; individuals that failed to germinate were replaced with similar genotypes to ensure a good sample size; plants were grown in controlled fields so the frequencies of the flower colors could be controlled; plants were planted in rows individually at regular intervals and equidistant from each other; plants were twined on their own bamboo stakes and not allowed to spread randomly on the ground or on each other (as they often do in the wild) in order to identify the flowers and seeds of individual plants; pests such as cutworms, deer, or other animals were kept out of the experimental plots. All the above were necessary to produce results with enough of a sample size to discern a pattern, and thanks to all of these I was able to finish a dissertation! And to be precise, I showed that under all the above conditions, balancing selection is important in its action. Yet because it is a reductionist science, I am left to wonder about the vagaries of everyday life, the random stochastic events that disrupt natural populations, the degree to which the design drove the conclusions. And there is always the matter of scale at which results emerge. Limitation of modern scientific life—graduate school, money, tenure clocks, resources, and weather patterns—all shape the knowledge the scientist, the context, and the culture of science and the plant co-produce.

In addition, my experiments with morning glories assumed a "genes versus environment" paradigm. Genetic lines of morning glory were formed to create uniform genetic contexts in which the only variant was the flower color at the w locus. The flowers were then grown in fields where little else grew. The experiments could not explore whether or not genes interact with other genes, or if the expression of genes at times depends on their environmental contexts in complex and even unpredictable ways. My carefully constructed experiments were not designed to capture the complexities of environmental contexts outside the experimental cycle. Latour's invocation of scientific experiments as theater is an apt analogy. The theater is itself built as an "objective" event with the scientist's own history and social identity deemed irrelevant, the experimental organism is stripped of all context and agency, and the objects of study are manufactured to minimize variation. The play itself is conducted on a stage that is carefully manipulated to produce as generic and sterile a location as possible so it can then be reproduced easily in other generic and sterile locations. With the observer, organism, and location set aside, the play unfolds to reveal some "truth" about biology and nature.

It is not that I contest the impulse for clarity that drives the research design and methods that I used in my training; it is more that this theater stands for pure knowledge about nature. Furthermore, so much is left out of the story of morning glories. I worry about the myopia produced by the rigid and futile effort to set boundaries on nature and culture that precludes and excludes so much complexity. How would the morning glory change if the boundaries were abandoned?

In many ways, I am arguing that most biological experiments (including my own) are, if we are precise, naturecultural experiments themselves. They are only deemed natural or rendered purely biological by ignoring the cultural contexts of theories, plants, and humans. Innumerable social and cultural assumptions are deeply embedded within the experiments and therefore in knowledge we produce about the natural world. Identifying these assumptions and reconceptualizing scientific practice as directed at the entanglement of nature and culture requires interdisciplinary effort. My primary argument is that if the biological sciences care about producing knowledge about nature and the humanities and social sciences care about producing knowledge about culture, both should be invested in the stories told about morning glories and other organisms. There was a historical and discipline-specific reason why certain questions and experiments seemed to me the most obvious in pondering a field of morning glory flowers. The central questions about variation that informed my research emerged, I was sorry to learn, from social/political debates about eugenics and "good" or "bad" traits. The experiments and methods I used were expressly designed to exclude a wide swath of information about morning glories over time and space in relation to humans' love/hate relationship to them. Understanding how the reductionist enterprise of science stripped the naturecultural contexts of science is a lesson in how we may imagine experiments that are more complex and interdisciplinary. Studying such naturecultural work tells us more about the biology of plants and their co-evolved histories with humans—the kind of reconstructive project worthy of the feminist studies of science.

CHAPTER TWO

A Genealogy of Variation

The Enduring Debate on Human Differences

It is not our differences that divide us. It is our inability
to recognize, accept, and celebrate those differences.

—Audre Lorde, *Our Dead Behind Us: Poems*

Shake your family tree and watch the nuts fall!

—Bumper Sticker

So we beat on, boats against the current, borne back
ceaselessly into the past.

—F. Scott Fitzgerald, *The Great Gatsby*

Elizabeth Wilson writes that we need to move away from those feminist discourses that "constitute the biological as fixed, locatable, and originary" (1998: 95). Rather, the biological is mutable, constantly changing, and evolving—both as a field and as the actual "matter" of biology. In this sense, a genealogical history of the idea of biological variation is a perfect case study of how the very idea of variation as the biological, locatable, originary fixedness is constantly shifting, reconceived, reinterpreted, and given new meaning. Starting from Charles Darwin, the idea of variation is a central organizing concept in evolutionary biology. Variation *is* the "stuff" of evolution, what we glorify as diversity or shun as evidence of mediocrity, and it has a genealogy of its own, strongly shaped by Darwin and his legacy of evolution by natural selection. It is this genealogy I trace in this chapter.

In connecting the idea of variation I encountered in evolutionary biology to ideas of diversity and difference in women's studies, I began by delving into the history of population genetics and evolutionary biology, and quickly encountered the specter of "eugenics."[1] With its contemporary connotations of holocaust, racial hygiene, genocide, and mass sterilizations of peoples considered "inferior," this is now a well-traveled history of the horrors of the unholy alliance of science and politics. Eugenics is often described as an "applied science" to improve the genetic composition of a population through controlling reproduction. Thus, bodies deemed inferior—whether due to sex, gender, race, ethnicity, sexuality, ability, nation, and the like—have historically been reproductively controlled, contained, or even exterminated. While eugenics has almost become interchangeable with racism, historically eugenic practices were broader, and in their global incarnations, often molded to serve national and local logics of class, caste, origins, geography, indigeneity, morality, and hierarchies of ability with categories such as "feebleminded" and "degeneracy" (P. Levine and Bashford 2010). Eugenics ranged from practices that prevented life (through sterilizations, contraception, segregation, or abortion), aimed for "fitter" lives (through practices to produce fitter children and families), generated "more" lives (through pronatalism, new reproductive technologies), or ended life (through euthanasia or nontreatment of bodies rendered inferior) (P. Levine and Bashford 2010, Lovett 2007). At its core, eugenics had an "evaluative" logic—deeming some lives worthier than others.

Eugenics was without doubt an important and fundamental aspect of many key movements in the past two centuries, intimately linked to ideologies of race, nation, and sex, and also a part of several programmatic and institutional initiatives, such as population control, social hygiene, state hospitals, and the welfare state (Dikötter 1998). Yet, as I got deeper into the literature, the picture became considerably more complex. This is an enormous literature, a veritable "eugenics industry" drawing on a wide range of historical materials (Pauly 1993). This history suggests, however, that we cannot write off eugenics as a few periods of excesses in the history of science. Much of the historiography has unfortunately focused on the extreme cases of race improvement in Germany, Britain, and the United States with less exploration of its global dimensions and also its wide appeal to a global populace across social, national, and ideological lines (Dikötter 1998). Rather, eugenics was often at the heart of well-meaning scientists' attempts to translate their biological research for human welfare. Diane Paul argues that "eugenics is a word with nasty connotations but an indeterminate meaning, indeed it often reveals more about its user's attitudes than the policies, practices, intentions or consequences labeled" (Paul 1995: 665–66). Contrary to the stereotype of the "sterile" scientists producing "pure" knowledge about the natural world, the history of evolutionary biology reveals a rich history of scientists concerned with "social problems" as they understood them. Eugenics thus was less about a clear set of scientific principles, and more about a "modern" way to discuss social problems in biological and scientific terms; politicians and scientists of various political persuasions appropriated eugenic discourse to promote social policy under the guise of objective and apolitical language of science and the laws of nature (Dikötter 1998, Bashford and Levine 2010). Thus, eugenics has had very different biological and political valances at different periods, embraced by an astonishing number and range of scientists with diverse political persuasions. Their goals were varied, as were their understandings of the relationship of biology, nature, and culture. To write off eugenics as "pseudoscientific," or as an unfortunate and embarrassing moment caused by a small number of scientists, ignores the wide support eugenics enjoyed and the involvement of many prominent scientists, and ultimately obscures the political nature of much of the biological and human sciences (Stepan 1982). But the easy story I began with—that the history of morning glory flowers and the shape of my dissertation were tied to a history of eugenics—was true but in far more complicated ways than I saw in the first hauntings among morning glory.

The history of eugenics is voluminous, indeed, as I came to discover, worthy of being a discipline in its own right. In my undergraduate and graduate courses in evolutionary biology, I learned about concepts utterly devoid of their political and cultural contexts. Here, I wish to argue that tracing the history of variation through a naturecultural framework reveals the inherent underlying logic of eugenics. A naturecultural framework allows us to see that evolutionary biologists have long wrestled with some version of what we recognize as the nature/nurture debates. Subsequently, the evolutionary theory, methods, and methodologies that they produced were entirely in service of their political goals and ambitions. Thus the idea of variation and its development in evolutionary biology is deeply implicated in political debates on biology and the social, and in the social policies of eugenics.

In tracing the idea of variation through the history of evolutionary biology, I adopt a Foucauldian approach in tracing a genealogy for several reasons. Variation is an idea, transcendent in many ways, which has been at the heart of both evolutionary biology and the politics of difference. A genealogical exploration allows us to see how this idea of variation—through mechanisms of power—shifts over time; how the veracity of its claims to truth and falsehood get decided; and how the idea of variation gets operationalized in the mechanisms of power and the state. This is not a search for origins, and there is no linear story. Rather, as Foucault suggests, this exploration reveals how power has fundamentally shaped truth claims. Truth, as I hope will become clear, is not easily apparent and "comes to be" through the complex workings of power. A history of the concept of variation in evolutionary biology is simultaneously also a history of biologists' quest for a "better" human society and the relationship of science and power. To trace the contours of power one must locate its meanings in its historical context. In each period, I show how the biological and social are not only intricately interconnected but indeed constitutive of each other.

In my view, the historical facts I outline have wide consensus and are non-controversial.[2] In chronicling this history, I deal with the major figures, including the four patriarchs, or fathers, of the field: Darwin, the father of evolution; Galton, the father of biometry; Malthus, the father of demography; and Mendel, the father of genetics. My key arguments are as follows:

- Eugenics is not a homogeneous movement or one with a common purpose or goals. Rather, eugenics is really an argument about the value and purpose of variation.
- Eugenic views were ultimately concerned with making a "good society." The political inclinations of the scientists shaped their vision of a good society and their views on the meaning and value of variation.
- Eugenics has shifting biological and political valances over time. An astonishing number and range of scientists of diverse political persuasions have embraced it.
- Eugenics is a critical site where scientific concerns about variation are converted into politically inflected meaning of diversity and difference.
- All evolutionary biologists engaged with eugenics as citizens, and their political views and values profoundly shaped their "doing" of science.
- The scope of genetics and that of eugenics are deeply intertwined, especially in the question on the value and meaning of variation.
- The meaning of variation remains unresolved and will continue to be a recurring question within science and society.

The Darwinian Revolution

While the idea of evolution has a much longer history, modern evolutionary biology emerges from the "Darwinian Revolution" (Ruse 2005, Paul 2001). It is here that the idea of variation emerges, and its importance is established. We begin with the publication of the *Origin of Species* in 1859. Darwin is critically important for three main reasons, as George Gaylord Simpson outlines. First, he brought together an immense wealth of evidence to support his theory of "descent with modification." Second, most biologists agree that the process of "natural selection" that Darwin advocated is the "primary force" of evolution. Finally, the evolutionary worldview that Darwin gave us is the *foundation* for later scientific theories. While there are interesting detours, all roads, including the ones leading to our morning glory experiments, lead to Darwin (Barlow 1995: 63–64).

Drawing heavily from observations of animal breeding, Darwin developed the idea of natural selection, a theory about "differential retention of heritable variation" (Lewontin 1978, Depew and Weber 1995: 5). Darwin's theory of adaptation also required an original theory of variation and inheritance—of transmutation and descent from a common ancestor (Depew and Weber 1995: 6), that is, he had to account for change over generations of evolution (Veuille 2010). To produce a dynamic theory of natural selection, Darwin named three key factors: phenotypic variation, differential fitness (differential adaptation to a given environment), and heredity (passing on these differentially fit characteristics) (Lewontin 1970, Grene 1990). Darwin's understanding and theories for each of these—*variation, fitness,* and *heredity*—have shifted considerably over time.

A *naturecultural* account of the evolution of Darwin's theories is anything but a linear or progressive story (Harwood 1989) for two main reasons. First, there emerged at many critical times in the history of evolutionary biology key debates that were not resolved easily or quickly. Second, Darwin's theories had profound implications for the human political world. Politicians, activists, thinkers (including key biologists in our story), and philosophers used these theories to advocate for and shape social policy. It is not that the biologists did biological work and then used these theories to shape social policy. Rather, it is in the interest of social policy that much of the biological work was done. One cannot separate the biology from the social—*the social is constitutive of the biology.* Others (including biologists) contested these theories and social policies. The shapes of the debates and history are largely contestations of social policy, that is, you cannot tell this story of evolutionary biology without the controversies in social policies.

Darwin and Our Conception of Variation

Exploring the social and political context of Darwin's work is an entire discipline in itself. Darwin's theories wrought not only a scientific revolution but also a cultural one. Michael Ruse argues that "Darwin's work came out of the culture of his day; it went back into the culture of his day" (Ruse 2005: 6). Two critical influences from classical economics significantly shaped Darwin's theory: Thomas Malthus and Adam Smith (Schweber 1978). While Malthus's theories were developed about human societies, Darwin extended and generalized his theory to all biological entities, arguing that there was competition for scarce resources and therefore those individuals that possessed traits that make them fitter or better adapted are likely to succeed; if these traits were heritable, natural selection would favor the better adapted traits. Adaptations are heritable traits that confer an advantage to individuals possessing them, and are built up through the process of natural selection over a number of generations, causing differential survival, or relative fitness (Brandon 1990, Depew and Weber 1995: 5, Lewontin 1978). Darwin brought Smith's idea of the "invisible hand" to biology by introducing the invisible hand of "natural selection" and creatively transferred the framework of a rational economy to the natural world—the idea that "free competition" is what built man was effectively transferred into the rational workings of nature (Gould 1979, Khalil 2000).

Thomas Malthus (1766–1834), founding father of demography (McCann 2009), has been well studied and his political motivations and sympathies are important and well known. Driven by a vision of markets and scarce resources, Malthus theorized the problems of growing populations with scarce resources. Malthus's theory of the "struggle in society" played a key role in leading Darwin to his own theory of evolution through the "struggle for existence" (Young 1990), popularized by Herbert Spencer's term *survival of the fittest*. The application of these principles to human betterment invoked ideas of "fit" and "unfit" individuals. As Mark Haller argues, early eugenicists had a clear solution to the dilemma of the growing "unfit." Through programs such as marriage restriction, sexual sterilization, and permanent custody of the defective, eugenicists hope to save society from the "burden of the unfit" (Haller 1984). You can see the direct embrace of these philosophies in Darwin's comment that "we Civilized men, on the other hand, do our utmost to check the process of elimination; we build asylums for the imbecile, the maimed, and the sick; we institute poor-laws; and our medical men exert their utmost skill to save the life of every one to the last moment . . . Thus the weak members of civilized societies propagate their kind. No one who has attended to breeding of domestic animals will doubt that this must be highly injurious to the race of man" (Darwin 1922: 136).

Darwin's study of Adam Smith and other economists stressed the "individual" as a central element and unit of his theory; his study of breeding suggested the possibility of random variation among individuals (Ghiselin 1971, Schweber 1978). He brought these ideas to bear in biology, arguing that even a complex organ such as the human eye could be "the supreme achievement of chance, operating under conditions of free competition and *laissez-faire*" (Keynes 1926). These ideological lenses are critical frames for Darwin's vision of evolution and also went on to undergird the social interpretations of Darwin.

While it took nearly eighty years for Darwin's work to be truly accepted (Mayr 1982, Depew 2010), Darwin profoundly transformed our views of variation. Historians argue that before Darwin, variation was largely viewed as imperfections of ideal God-given "types." Darwin shifted our conception from typological thinking to variation as the raw material for natural selection, an emphasis on concrete things (individuals) delegitimizing both essentialist and teleological ways of thinking (Ghiselin 2005). Ernst Mayr characterizes this as a shift from typological thinking to population thinking (Mayr 1976, Sober 1980).[3] Marjorie Grene wittily summarizes: "It is an article of faith in evolutionary biology that Darwin shifted biological theory and biological research from a static, evil, 'metaphysical' kind of thinking called essentialism or 'typology' to a dynamic, good, scientific conceptual style called 'population thinking'" (Grene 1990: 237).

Population thinkers do not see variation as imperfection; they do not ask which archetype an organism resembles, but rather *how* it differs, ever so little from its neighbors. In this sense, populations are seen as variegated collections of differing individuals—there is nothing normal/typical of any individual. Population thinking thus tracks changes in the aggregates of individual characteristics, focusing on frequency changes. The core of contemporary Darwinism identifies several key steps for a descent with modification. First, variation arises randomly with respect to the needs of the organism. Second, the importance of variation depends on a given environment. Some variation confers advantage to an organism, other forms of variation are harmful. This gives rise to differential adaptive advantages. Better adapted individuals (by definition) will have more favorable reproductive outcomes, leading to change in frequencies—it is these shifts that lead to evolutionary change. Evolution is in that sense "exoteleological," that is, extinction comes from outside and therefore there is always an ecological basis to evolutionary theory.

Thinking through eugenics clarifies how population thinking and typological thinking get inextricably interconnected. It is a paradox that Darwinian evolution is seen as moving away from typology to population thinking, while in practice evolutionary biologists focused on the "aggregates" of individuals through time. A field that focused on individuals—drew generalizations to groups (group inferiority)—and then social policy worked to police individuals in those groups. The consistent conflation of essentialism at the population versus individual level (Sober 1980) gives eugenic thinking its power. Despite the apparent move from typology, in practice "types" are very apparent in how biological knowledge gets applied in the field—from Galton's "types" to the contemporary resurgence in discourses on "race" or "populations."

Darwin's biggest problem was an inadequate theory of inheritance. While he proposed blending inheritance as a mechanism, it did not work well with his evolutionary theories. It was only after Mendel's work was rediscovered in 1900 that the ideas of variation and inheritance were reconciled into a cohesive, functioning modern evolutionary synthesis.

Galton and the Origin of Eugenics

The influence of Darwin's theories is best understood through his cousin, Francis Galton (1822–1911). Galton's greatest legacy is in his attempt to strengthen (by the data he gathered and the biometrical models and statistics he developed) as well as apply Darwin's ideas to human societies (in his theories and recommendation of eugenics). A formidable intellect in so many disciplines, Galton was visionary in this sense—seeing the importance of Darwinian evolution in its application to human betterment through selective breeding. Contrary to the horrors that the term *eugenics* evokes today, Galton saw eugenics as humane, arguing that it would ensure "natural selection by other processes that are more merciful and not less effective. This is precisely the aim of eugenics." Galton's eugenics was informed by the logic of Darwinism fused with a scientific narrative on improving society—this scientifically grounded legacy is Galton's profound contribution, a legacy that lives on to this day. In addition to *eugenics,* Galton also coined another famous phrase, *nature and nurture.* Galton had a clear position in the debate. He was convinced that it was nature and not nurture that determined hereditary ability. He studied the heritability of fitness, what he called "talent and character," by using pedigrees, twin studies, and anthropometric measurements (Gilham 2001). In the process of this work, he invented the field of biometry and key concepts in statistics—such regressions, correlations, and popularizing the use of the normal distribution. Indeed, I used several of these statistical tools in my work on morning glories.

The Biometrican/Mendelian Controversy

Almost contemporaneous with Darwin, Gregor Mendel in a series of important experiments during 1856–63 developed a theory of inheritance. Unfortunately, his theories went unnoticed until independently rediscovered by three European scientists in 1900. It was not immediately obvious how Mendel's theory of inheritance would augment Darwin's theory of natural selection. A theory of evolution was widely believed, but the mechanism was still in dispute. The attempt to resolve the nature of variation, its mechanism, inheritance, and evolution, would take another decade or two through a controversy during 1900–1915 between biometricians such as Karl Pearson and Raphael Weldon, and Mendelians such as Gregory Bateson and Hugo de Vries.

Galton's work produced two schools of rival theorists (Marais 1974), who disagreed on the nature of variation—was it continuous or discontinuous? The idea of discontinuous variation, where evolution occurred in saltations, or jumps, meshed with the Mendelians. But many traits Galton described were best conceptualized as continuous, in a normal distribution with population means and variances. Selection could thus shift the mean over many generations (Dietrich 2006). The debate raged on in scientific journals and outside, at times with considerable personal animus. The growing data on both sides saw a resolution of Mendelian and biometrical approaches (Kim 1994, Dietrich 2006).

During the period of discussion here, the late nineteenth century and the early decades of the twentieth century, educated classes of England were primed to welcome eugenics. Gilham suggests two main reasons. First, amid lower birth rates of upper and middle classes and higher birth rates of the working poor, eugenic ideals thrived. Why not support policies limiting the fertility of the poor while encouraging those of the elite, thereby improving the national "stock"? Second, by then acceptance of Darwinism was growing. Such arguments had widespread appeal not only in England but also in much of Europe and the United States. Victorians worried that natural selection was no longer at work because modern medicine, charity, and other humanitarian measures prevented the elimination of physically and mentally deficient traits (Paul 2001).

In the United States as well, eugenics established deep roots among the privileged. In 1906, the Race Betterment Foundation was founded, and in 1910, the Eugenics Record Office (Allen 1986). Eugenicists worked hard to successfully transform both public policy and popular culture. These organizations had wide appeal, bringing together scientists, academics, museums, clergy, and philanthropic organizations. Newsletters such as the *Eugenical News* spread the message of eugenics: regardless of context, immigrants, persons of color, and social misfits were seen as unfit (Selden 2005). Immigration and sterilization legislation is a testament to their success (Lombardo 2002).

The new emerging biology profoundly shaped social thinking. David Star Jordan, president of Stanford, declared in 1914 that "good stock" was necessary for the production of good children (Cravens 1978: 46). The rhetoric of early eugenics was a rather crude form of biological determinism, a broad stroke against the biological fitness of whole groups such as immigrants, communities of color, poor, and marginalized (Selden 2005, Kluchin 2009). Indeed, eugenicists believed that poverty, criminality, illegitimacy, epilepsy, feeblemindedness, and alcoholism were all traits that could not be altered (Kluchin 2009). Using overly simplistic views of human development, eugenicists successfully transformed these ideas into coercive and regressive social policies to enact state control of immigrants, communities of color, and women's bodies (Selden 2005). In the technologies of biopower, women, seen primarily as reproductive agents, emerge as the key site of the struggle against difference (Mottier 2008). Women were also key players in the infrastructure of eugenic enforcement by promoting eugenic ideologies through women's groups, churches, and charity (Haller 1984). The corollary, of course, is always true—the "fit" and superior emphasize the "natural" aptitude of white, native-born Americans, and these policies consolidated their social, economic, and political power (Lovett 2007, Kluchen 2009). Eugenic measures had worldwide appeal, reaching practically across the globe (P. Levine and Bashford 2010). Rather than a wholesale export to the rest of the world, however, eugenics was transformed to suit individual contexts and used to reinforce local hierarchies of power (Ramusack 1989, Anandhi 1998, Dikötter 1998, McCann 2009, Klausen and Bashford 2010, Hodges 2008, 2010).

The biometricians and Mendelians were deeply embedded in eugenic rhetoric and action. It is the scientific patina that propelled the rigor and certitude of eugenic thinking. Men of science such as Galton and Pearson popularized eugenic ideals as logical conclusions of sound science. What emerges in reading about their lives is how deeply implicated each facet of their lives—personal, political, and scientific—were with each other. For example, Nicholas Gilham (2001) in his excellent biography of Francis Galton argues that eugenics was fundamental not only to Galton's political work, but also his scientific work. Galton, he argues, saw his work as entirely moving toward a theory of eugenics.

In his biography, Theodore Porter argues that Pearson became the "harshest of social Darwinism, a champion of neo-Darwinian selection" (Porter 2005: 214). Reproduction became social regulation. If for Galton the new eugenic society was to be run by "eugenic priests," then for Pearson it was to be run by the new aristocracy of merit, supported by eugenic scientists (Porter 2005). Pearson moved away from ordinary mathematics into biometry precisely because it allowed him to create a truly scientific form of social Darwinism (Norton 1978). Pearson was not so much a social Darwinist as a socialist Darwinist, however. To him, reproduction should be regulated by science, not by poverty or hunger.

He called for accessible technical and commercial teaching for all, arguing that the future depended on fitness and health of working people (Porter 2005). In contrast to Pearson and Galton, Bateson, a Mendelian, was scornful of eugenics. Here again, historians suggest that his scientific views were shaped by his political and economic views. With his strong hatred of utilitarianism, they suggest that Bateson passionately disagreed with the ways in which the biometricians connected a gradualist and utilitarian philosophy of nature with an economic philosophy he abhorred (MacKenzie 1981, Norton 1983).

Biometrics was "generated as a toolbox for eugenics, as a collection of measurements of characteristics and correlations for the analysis of inheritance" (Louçã, 2008: 255). The foundations of eugenic thinking, its methodology, theories, and statistical modeling, are deeply indebted to biometry. "It is an inescapable part of Darwinism's history that statistical Darwinism, in both its progenetic and genetic form, was for a long time an instrument in the service of the eugenics movement" (Louçã 2008: 263). Biometrics was deeply rooted in eugenics and eugenics was supported by the majority of the profession.

"Eugenics was hegemonic in the universe of ideas among this population of scientists by the turn of the century" (Louçã 2008: 280). Theories and results from plant and animal studies were extended to humans (Cooke 1997). Eugenics was also a culture, providing a bridge between different fields, exerting a powerful influence for defining an agenda for biology and the social sciences where they would inform each other. "Statistics was the language used for that communication and it was a new language—society and not only individuals mattered" (Desrosières 1998: 68, Louçã 2008).

The study of variation shaped some of the most important tools of statistics. In the late nineteenth century, important statistical innovations "came from fields in which statistics was used to model variation not just to estimate or reduce error" (Gigerenzer et al. 1989: 68, Louçã 2008). Modern statistics thus integrated scientific and administrative practices (Desrosières 1998). Indeed, the nature of modeling variation and evolutionary change was at the heart of the resolution of the controversies (Louçã 2008). Like Darwin, Galton and Pearson had wide-reaching impact. The scientific "truth" of eugenic doctrine came to dominate economics, shifting economics from assumptions of human homogeneity in the classical period to ideas of foundational differences among and within races of people in postclassical economics (Darity 1995, Levy and Peart 2004). The growth of eugenics in the twentieth century is closely intertwined with the ascendency of the power of science as science battled traditional forces such as religion for the heart of modernity (Turda 2010).

An Evolutionary Synthesis

The modern evolutionary synthesis resolved the biometrician/Mendelian debate about the nature of variation and its inheritance. The resolution involved two questions: first, whether genetic variation was continuous or discontinuous and second, whether evolution was gradual or saltational. The synthesis produced over a decade (1936–47) resolved the controversy by showing that Mendelian genetics was consistent with natural selection and gradual evolution. A new synthesis with a sophisticated mathematical framework came into being. A critical byproduct of the biometrician/Mendelian controversy was the important statistical and mathematical methods formalized in the early twentieth century that have remained foundational for the field. What emerged was a mathematical population genetics usually associated with Sewall Wright, R. A. Fisher, and J. B. S. Haldane. Their work set the foundations of population genetics, which formally reconciled Mendel and Darwin (Provine 1971, Dietrich 2006). The synthesis defined the emerging field of evolutionary biology by explaining and incorporating the many subdisciplines and branches of biology such as genetics, cytology, systematics, botany, zoology, morphology, ecology, and paleontology, as well as mathematics and statistics (Dietrich 2006, Smocovitis 1992, Grene 1983). Developing the synthesis involved, as Provine argues, "a vast cut-down of the variables considered important to the evolutionary process" (1988: 61). More important, the field concluded that there was no purposive force behind evolution (Provine 1988, Dietrich 2006).

For the first forty years of the century, genetics was synonymous with eugenics (Mazumdar 1992). "Eugenics was the dog that wagged the tail of population genetics and evolutionary theory, not the other way," argues Norton (1983: 21). In understanding scientists of this period, eugenics is key (Norton 1983, MacKenzie 1981, Searle 1976). Key figures in science continued the tradition of being firmly grounded in politics and society. Eugenics was offered as a scientific solution by some, a technocratic solution by others, and as social engineering by yet others. Many of the leading scientific figures of many political stripes were eugenicists. While the motivations, goals, and ideological views of scientists varied greatly, the idea of "human betterment" through science had a large following. Eugenics is ultimately an ideology fundamentally about using science and social policy to produce a better human society, deployed by the political right and left (Paul 2001). Eugenics is not one thing, and had wide-reaching ideologies and politics; not all eugenics was put to use for racist, classist, and sexist ends; some were purposefully constructed to fight inequality. Scientists' vision of a better human society directly shaped their theories of variation, its importance and its application to human societies. "Eugenics," L. C. Dunn wrote, "has come to mean an effort to foster a program of social improvement rather than an effort to discover fact" (qtd. in Gromley 2009: 43).

In contrast, of all the key figures, R. A. Fisher was perhaps the most ardent of the eugenicists. Firmly grounded in eugenic ideology, much of his mathematical and evolutionary work was developed in concert with and perhaps in the cause of an evolutionary theory that explicated eugenic ideology. One-third of Fisher's foundational text *Genetical Theory of Natural Selection* connects the mathematical and theoretical arguments to human populations and is devoted to eugenics (Gould 1991, Norton 1983, Moore 2007). Here, Fisher's genetic explanations of abstract populations inspires his general theory of racial decline (and possible eugenic salvation) and his belief that advanced civilizations destroy themselves by "the social promotion of the relatively infertile." In a wonderful essay on R. A. Fisher, Stephen Jay Gould remarks, "We don't like to admit flaws in our saints" (Gould 1991). As biologists, we rarely acknowledge them or make note of them; they are largely absent from our internalist histories and at best they are seen as an "unfortunate and discardable appendage," even if they appear in "our profession's bible." We need to understand the eugenic work as central to their science and cannot choose to ignore it as "an embarrassment" (Gould 1991).

The eugenic motivations of the key scientists seem so apparent and indeed so instrumental to their scientific views and lives that it is hard to write off their views of politics and science as a benign dalliance or side project. It is indeed interesting that internalist histories of evolutionary biology largely sideline a discussion of eugenics or present a benign picture of the major figures. Instead, most were eugenic enthusiasts (Paul 2001). Some like Dobzhansky and Haldane came from the political left and saw the utopian potential of a new experimental biology that could predict and manipulate biology to enhance human futures. For Haldane and other scientists who bought into these promises, the first decades of the twentieth century were exciting, promising human improvement on many fronts. Therefore, "it was still possible to be a scientist, a socialist, a meritocrat and a eugenicist" (Tredoux 2001). As Mendelian inheritance laws helped develop novel techniques in plant breeding and revolutionized agricultural breeds and yields, so biologists saw the potential for human improvement (Mark Adams 2000). This new future-oriented biology would replace the old religions.

Historians also challenge the contention that there was a profound shift after World War II that undermined eugenic thinking and ideologies of "race hygiene" (Gannett 2001, Reardon 2005). Race was not relegated to the "scrap heap" of history with the advent of modern evolutionary biology, but rather redefined (Reardon 2005). Instead, Gannett argues that races were reconceptualized as populations; a populational concept of race replaced a typological one. Indeed, the concept of race has made a resurgence in contemporary biology (Bliss 2012, Reardon 2005, Roberts 2012).

Women and gender are central to eugenic practices. Eugenic practices were characterized by rigid sex and gender ideologies and enforced through top-down reproductive control of women's bodies (Stern 2010). Feminists remind us that to understand eugenics, frameworks of women's and gender studies are essential. These practices cultivated eugenic practices through promoting particular ideologies of femininity and motherhood, stratification of reproductive health technologies, and access to public health (Stern 2005). As always, these ideologies were implemented strategically across the globe.

Classical/Balance and Neutralist/Selectionist Controversy

Once the evolutionary synthesis developed a robust theory of the nature of variation and its inheritance, modern molecular techniques unearthed an astonishing amount of genetic variation at the molecular level and added a whole new perspective on the genome. Is variation good or bad? As we saw in the previous chapter, the classical/balance controversy that raged from 1950 to 1970 dealt with the question of the nature and maintenance of variation (Lewontin 1987, Dietrich 2006).

What does this variation mean? What was the relationship between genetic variation and natural selection? As we also saw in the previous chapter, the opposed classical and balance positions were promoted by Muller and Dobzhansky (Beatty 1987, Dietrich 2006). Again for both, their genetic theories of variation were inextricably linked to their ideological views on variation, diversity, and eugenics. To Muller, all mutation was deleterious and thus an active eugenics program was necessary to foster the purifying effects of natural selection. In contrast, Dobzhansky saw variation as generally adaptive and thus the presence of variation was a sign that natural selection actively maintained genetic variation through balancing selection. The controversy, however, was a stalemate. Rather than resolve itself, the controversy was transformed into a new question. Perhaps most of the variation was neutral, rather than beneficial or deleterious? With the discovery of the structure of DNA as a double helix, molecular data exploded, leading to the neutralist-selectionist controversy (Dietrich 2006).

The controversy resolved in the 1980s with the advent of DNA sequencing. New statistical techniques allowed geneticists to detect selection at the molecular level (Kreitman 2000). While earlier tests could not tell the difference between neutrality and selection, these tests could. Thus neutrality became the null hypothesis and a way to look for whether selection was in action (Dietrich 2006). It is in this tradition that the morning glory experiments were structured.

Through all these controversies and debates, the field of evolutionary biology came into its own, and continued an uneasy relationship with its roots in eugenics and biometry. Fields such as animal behavior, sociobiology, and evolutionary psychology translated results from plants and animals to ground evolutionary arguments for human populations. The debates on nature and nurture rage on. These innovations in evolutionary biology profoundly shaped the emerging fields of human population biology and demography and subsequently molecular biology and medical genetics.

Variation in Human Populations

The biggest recognition after the war was the horrors of eugenics, especially in its links to race. Despite scientists distancing themselves from the simplistic biological determinism that underlay eugenics, contentious debates around the ideas of race, population, and diversity continued (Roll-Hansen 2010). The same fault lines emerged around "ideological approaches to the study of human diversity and rigorous scientific ones" (Reardon 2005: 34). For example, are races discrete units or "open genetic systems," that is, not "discrete and qualitative" but "overlapping and quantitative" (Reardon 2005: 34)? This basic question of race and human variation reemerges in contemporary debates on human diversity (Reardon 2005: 35). It is no surprise that this superficial repackaging of race is easily reversed and that race resurfaces in biological thinking today.

Tracking the idea of variation in the postwar years largely takes us to the growing field of demography, which continued to have links with evolutionary biology and population genetics. The field of demography demonstrates how through quantification practices the concept of population itself "constitutes a technology of liberal statecraft" to govern a nation of "autonomous individuals" (McCann 2009: 144). Population science owed its debts to population biology. The Darwinian or eugenic paradox was the central problem in eugenic thinking—social success was negatively correlated with biological success (Kevles 1985, Vining 1983: 103). Why?

In the postwar years, in an attempt to distance themselves from the stigma of eugenics, American demographers increasingly attempted to define demography as a social science, and less as a biologically based eugenics (Ramsden 2009). Relations between biologists and social scientists grew increasingly tense (Ramsden 2009). Are patterns of population growth a biological problem caused by the decline of fertility among the rich, or a sociological culmination of the process of urbanization and industrialization, accompanied by the increased use of birth control? In the postwar years, the eugenics movement pursued a "simultaneous program of withdrawal and expansion." In the 1950s U.S. demographers were "*the* scientific arbiters of population science" who were responsible for establishing international standards through the United Nations Population Division (McCann 2009: 142). In the 1960s a revitalized American Eugenics Society helped reunite leading demographers and geneticists.

Thus began the era of "overpopulation," and global programs sought to stabilize populations through seemingly benign frames of "family planning." The crisis of population "explosion" was grounded in familiar Malthusian logic that undergirded Darwinian thought. While the focus was more about population size or quantity rather than quality, the logic of population control focused on controlling "some" populations more than others was reminiscent of earlier eugenic logics. Within the developed worlds, sterilization and contraceptive practices targeted communities of color, immigrants, indigenous populations, and the differently abled (D. Roberts 1997). Within a global context, highly funded projects worked to control the population of third world countries (E. Hartmann 1995). Within third world contexts, the ideology of population control became synonymous with modernity and progress. Thus eugenics remains an important part of the history of the "modern subject," especially the "modern liberal subject" whose individual rights to factors such as reproduction and health were tied to national logics of population. This is most apparent through population control policies of contraception, sterilization, and abortion (P. Levine and Bashford 2010). The emergence of genetic demography signaled an era where geneticists and demographers tried to revitalize social and biological scientists within an interdisciplinary program allowing for eugenic improvement consistent with welfare democracy (Ramsden 2006).

Throughout this rapprochement of genetics and demography, scientists were key players and proponents of eugenic ideas. The popular argument that scientists entered the "population debates" and purified it of politics is not supported by the historical record (Hodgson 1991). In the end, demographers viewed genetics not as a threat, but as an ally in the development of their discipline. For geneticists, the collaboration allowed them to engage more fully and effectively in the field of human evolution. Thus, the eugenics movement played a crucial role in bringing geneticists and demographers together (Ramsden 2008).

Eugenics by Another Name? A Return to the Nature and Meaning of Variation

Hansen et al. (2008: 104) maintain that "two hundred years after the birth of Charles Darwin, his theory of natural selection continues to inform current practice in medicine and the related discipline of bioethics or health ethics . . . We contend that there is a connecting thread of essentialist Darwinism that links contemporary eugenic practices with a long, and often disturbing history." From 1980 to the present, two main strains of earlier eugenic practices continue—practices to control population growth and human health (through medical genetics). These shifts, however, have happened alongside profound changes in world governments and political ideologies. National governments have increasingly retreated from their roles in the welfare of their citizens, and centralized state logic and governance have been replaced by market logic and individual "choice" and decision making (Lemke 2007). These shifts are apparent in the transformations in both population control and medicine. In the former colonies or the third world, family planning efforts continue, and now include mechanisms of the market and "individual" choice and control. Similarly within the United States, welfare reform under neoliberalism has dramatically shifted the economic, governmental, political, and public policy contexts of reproductive policies (D. Roberts 2009). Yet the end results are the same: economic, racial, and politically marginal communities continue to be targets of reproductive control. Human health and reproduction has also shifted. Rather than state-sponsored laws or rules that determine which lives are "worth" living, decisions are now left to individual "choice" and decision making. For example, statistics suggest that as many as 90 percent of women presented with a positive prenatal diagnosis of Down Syndrome choose to abort the fetus (Harmon 2007). This shift from state policy to a market-driven individual model is what marks the profound changes that have occurred. In both cases, I argue, the end result or the logic of which bodies are rendered desirable and worth living, which bodies are controlled or eliminated, remains the same. Since I have discussed the demographic turn earlier, I focus here on the burgeoning field of molecular biology and genetics.

In "Is a New Eugenics Afoot?" Garland Allen, reflecting on the horrors and logics of Nazi racial science, argues that "we are poised at the threshold of a similar period in our own history and are adopting a similar mind frame as our predecessors" (Allen 2001: 61). Allen makes a compelling case for a visible resurgence in such logics of efficiency, economic hardship, and human betterment and improvement. Genetic counseling, pre-genetic diagnoses, preferential abortions, and talk of genetic enhancements and designer babies are all seen as modern incarnations of eugenics (Hansen et al. 2008).[4] Yet, the "new genetics"[5] is vigorously defended as being different from the "old eugenics" (Kerr 1998, Hubbard and Wald 1999).

The old eugenics is characterized as founded on racism, an "outdated" and "discredited" ideology of the past. In contrast, the new genetics is seen as divorced from this tortured history. With the growing geneticization of our culture, however, genetics has come to stand in for a "set of ideals about a perfect health culture" and is thus argued to represent "modern day eugenics" (Nelkin and Lindee 1995: 191). Some of the distinctions made are that while the old eugenics was predicated on coercive action, the new genetics is "voluntary," and the shift from "good of society" to "individual rights" is significant (Paul 1995). In contrast to the earlier focus on "heredity and behavior," the new geneticists recognize "heredity *and* environmental influences." Furthermore in contrast to altering gene pools, the new genetics is argued to diagnose and treat diseases (Kerr 1998). But, as Foucault argues, claims of the decentralized voluntary nature of biopower is one of the most striking features of modern governmentality. Some of the core debates that animate contemporary genetics and evolutionary biology retain their historical legacies. We see the old debates live on in current times through the following:

- *Questions of Normalcy:* As Hansen et al. (2008) argues, Darwin's theory of natural selection continues to inform current medicine and bioethics/health ethics. For example, questions of impairment/disability are usually not seen as "natural variation" but rather as abnormal and deviant. Technological fixes to normality are ubiquitous (Martin 2006, Dumit 2012).
- *Biological Determinism:* Biological determinism thrives with claims of the genetic basis of diabetes, hypertension, homosexuality, math ability, and religion! Most human characters have been geneticized. The old debates on IQ live on as potent reminders of an ideological battle of nature versus nurture that rages on.
- *The Resurgence of Race:* Despite the 1950s declaration against the biological basis of race, race lives on as a potent biological and social marker. Biologists continue to use race as a category of analysis (Gannett 2001, Reardon 2005, Duster 2003, Bliss 2012).
- *HapMaps:* With the advent of the Human Genome Project, individuals are being sequenced across the globe. Haplotype maps (HapMaps) map global human genetic patterns with underlying claims of the ge-

netic basis of health and diseases. BiDil, the first ever "ethnic" drug, was approved by the FDA as recently as 2005.

- *Personalized Medicine:* The resurgence of race and targeted medicine are all part of the growing move to the biologilization of life (Kaufman and Morgan 2005, H. Rose 2007). The ultimate claims of these projects are that we will one day have individualized, personalized medicine—all derived from our individual DNA.
- *Overpopulation Policies:* The rhetoric of overpopulation continues. As always, the solutions continue to be about controlling the fertility of women (Hartmann 2011) The rhetoric in much of the west—anxieties of the lower fertility of the elite and the high fertility of the poor and foreign (the so-called eugenic paradox)—continues to shape the same eugenic thinking today.
- *The New Reproductive Technologies:* Nowhere is eugenic thinking more readily apparent than in the new reproductive technologies. Almost exclusively in the hands of the rich, reproductive technologies have pro-liferated. While reproductive technologies offered liberatory potential, they have also reinscribed the narratives of the biological and genetic connections with the offspring.
- *Epigenetics:* One of the exciting subfields in genetics today is that of epi-genetics—studying heritable change in the phenotype that are caused by mechanisms other than a change to DNA sequences. Epigenetics suggests that nongenetic factors such as methylation, chromatin remod-eling, and RNA transcripts can cause an organism's genes to behave differently. These changes last not only for the duration of a cell's life, but can linger for multiple generations. On the one hand, this work has radical implications—the story of Darwinian evolution and genetics entirely retold. The separation of genes and their environment has been such a cornerstone of the evolutionary narrative that the possibility that an almost Lamarckian narrative of the inheritance of acquired charac-ters is astonishing and destabilizing. Yet, in practice what appears to be happening is that claims of the power of the environment are being translated into rigid and regimented regulation of women's pregnant bodies. Pregnant bodies and motherhood have become a potent site of mobilizing our dreams of a future "good" society (Waggoner 2012).

The 1990s saw the resurgence in arguments that eugenics is not a bad thing (Roll-Hansen 2010, Caplan et al. 1999) and talk of an "inescapable eugenics" (Kitcher 1997). Some unapologetically and openly call for a return to eugen-ics. To feminist, critical race theory, and disability rights scholars who study populations on whom eugenic policies have largely been enacted, eugenics has a continuous and unbroken history. Merryn Ekberg argues that "the old eugenics was genetics and the new genetics is eugenics" (Ekberg 2007: 591). Troy Duster argues that contemporary genetic technologies and policies lead us to the same house of eugenics but through the backdoor (Duster 2003). The core of the debate is about intent since the "new genetics" is not *applied with eugenic purpose* (Roll-Hansen 2010). The shift from state policy to personal choice is a shift from coercion to volunteerism, from population to individual, and from public to private practices (Bashford 2010). This mischaracterizes the issue, however, by overstating the freedoms of contemporary genetics or the easy separation of individual versus population (Bashford 2010). Individual choices by parents still create population-level changes (Rolls-Hansen 2010).

Indeed, in the twenty-first century, deliberation and decision making have themselves become technoscientific objects and any idea of choice for the parents is fiction (Samerski 2009). To some, the logic of genetic counseling encodes centrally a eugenic logic and is "social engineering" (Samerski 2009). Is this an exercise in autonomy, or a new duty to manage oneself according to the rules of neoliberalism (Novas and Rose 2000, H. Rose 2007)? Does it matter that health "has come to represent for the neoliberal individual who has 'chosen' it, an 'objective' witness to his or her suitability in function as a free and rational agent" (Greco 1993: 369–70)? To others this new formation is coercive, but is not eugenics (H. Rose 2007, Cowan 2008). As Rabinow and Rose remind us, something has changed since "*letting die* is not *making die*" (Rabinow and Rose 2006: 211).

Thus definitions of eugenics continue to be debated. Hilary Rose argues that the twentieth century was "the century of eugenics, for genetics and eugenics have like conjoint twins, both individual and linked histories over the course of that one hundred years" (2007). Perhaps it is more accurate to say that there was no single point of culmination or transformation in the history of eugenics; rather, we are neither at the beginning nor the end, but in the middle (Rose 2007). In the varied formations of the complex relationships of science and politics in eugenics, what we mean by eugenics is always contested and shifting, not only across time but in fact in every historical period.

Engineering the Human: Legacies of Eugenics

In his famous 1973 essay of the same name, Theodisius Dobzhansky once famously quipped that "Nothing in biology makes sense except in the light of evolution." At least since the time of Darwin, evolutionary biology has been an important if not central organizing force in the biological sciences. Galton coined the term *eugenics* in 1883 and since then the term has come to have a life of its own. What I hope is clear is that eugenics is a much more heterogeneous, complex, and nuanced category than popular conceptions accord it. Diane Paul rightly suggests that it is time for us to develop a more sophisticated account of eugenics, "not for the sake of fidelity to the historical record but for a more adequate public policy . . . to suggest that there is a historical connection to eugenics is quite different from arguing that it is eugenics" (Paul 2007: 15).

A naturecultural genealogy of variation in evolutionary biology is especially helpful. In wrestling with the idea of variation, we see how the scientific knowledge of each period was profoundly shaped by political questions. Earlier efforts to understand diversity through chromosome mapping and blood linkage studies have transitioned into modern projects of studying diversity through genetic mapping. This is a continuous history in both institutional and intellectual realms (Mazumdar 2002). In discussing the British case, Mazumdar argues that human genetics was the successor of eugenics, so when the British Society finally wound down in 1989, it "changed its name and moved out of town, leaving the field to human genetics" (Mazumdar 1991: 257). Fisher's formulations of genotypic and phenotypic variation opened a direct path to the Jensenist debates on IQ in the 1970s that revisited the environment and heredity question (Norton 1983, Kamin 1974). The terms of the debates on race and human variation in the postwar era have now reemerged in contemporary debates on human diversity. Reardon suggests that in historical periods such as the case of Nazi science where there exist "states of domination," liberal efforts to extend powers to research subjects may be important. In the absence of such domination, however, the attempt to "genomic liberalism" may in fact create new modes of racism precisely when scientists seek antiracist modes of inquiry (Reardon 2011).

To suggest a continuous history is not to suggest that nothing has changed. We are not merely repeating history. Yet the history of variation has an unmistakable déjà vu quality because in the past century and a half while we have developed a theory of variation, its nature, and its inheritance, the question of the *meaning* of variation for human society remains unresolved. Eugenics in its myriad manifestations and theories of variation all seem intertwined with changing theories of the relationship of nature and nurture, genotype and phenotype.

Living with Ghosts: Lessons from a Eugenic Past

Ghosts produce material effects. "To be haunted in the name of a will to heal," Avery Gordon (1997: 57) argues, "is to allow the ghost to help you imagine what was lost and never even existed, really. That is its utopian grace: to encourage a steeling sorrow laced with delight for what we lost and we never had; to long for the insight of that moment in which we recognize, as in Benjamin's profane illumination, that it could have been and can be otherwise." In tracking ghosts, scientific practice transforms into a deep-seated historical practice, where the objects and subjects of science and their histories come hurtling into focus. But it is this kind of seeing that disciplinary thinking has systematically excluded. In historicizing the study of morning glory flower color variation, the eugenic histories of variation and theories of natural selection emerge. It is in tracking these ghosts—the women, the men, the "other," of differing nationalities, classes, ethnicities, sexualities—that the violence of the past and present comes starkly into focus. It is these ghosts that live within us and within the specters of the natural and cultural.

So, what does this perusal of a genealogy of variation tell us about the experiments I began this section with? The histories are embedded in the project: the frame of variation, the problem of variation, a theory of heredity, mechanisms of natural selection, balancing selection, heterosis, all draw from this history. It is no wonder that I looked at a field of morning glories and asked why there were so many colors. Throughout my scientific education, there was little emphasis on the eugenic preoccupations of these so-called forefathers. From my naturecultural perch, I now see the ghosts wandering the morning glory fields. Gory, bloody, violent ghosts from our eugenic past.

Two other things are striking, too. First, as we saw in the previous chapter, reductionist science decontextualizes, homogenizes, and purifies its objects of inquiry. The whole point of the experiment was to homogenize the context of the plant—its genetics, ecology, and history. Indeed, morning glories have a fascinating social history, but like humans, there is no place for the agency and histories of plants in traditional science. Second, scientific inquiry and training is structured precisely around the erasures of the contexts of discovery. Much of evolutionary biology was built around eugenic goals, its champions, and ardent biological and social activists. It is repeatedly clear that the science was produced with the aid of eugenics; eugenics is constitutive of the science. Yet, science ignores this history and deems its intellectual and political roots irrelevant. What is at stake in this erasure?

What is striking is how enduring Darwin's and Galton's frameworks remain. Despite recent developments in molecular biology and genetics and the molecularization of health, we continue to be haunted by the familiar nature/nurture debates. How do we explain diabetes and hypertension? Genes or environment? As with earlier times, these are deeply political questions and scientists today debate the same questions although with a very different understanding of biology and genetics. But there is something even deeper in this sense of déjà vu and the hauntings. We began with Darwin and the question of the nature of variation and inheritance. It was not until the modern synthesis that Darwin's theory of natural selection was rendered compatible with a Mendelian theory of inheritance. Yet, the field of epigenetics has now brought this familiar story back into question. Resurrecting Lamarckian ideas, the field of epigenetics reinvokes questions of the mechanisms of inheritance that haunted an earlier era. Perhaps the environment can shape the genome and change the expression of genes in future generations?

Avery Gordon, drawing from Zora Neale Hurston, writes,

> Ghosts hate new things . . . The reason why is because ghosts are characteristically attached to the events, things, and places that produced them in the first place; by nature they are haunting reminders of lingering trouble. Ghosts hate new things precisely because once the conditions that call them up and keep them alive have been removed, their reason for being and their power to haunt are severely restricted. (1997: xix)

After 150 years of Darwinian thought, we have not outgrown the basic binary formulations of nature/nurture or genes/environment. The recursive nature of this history is striking. Indeed, the ghosts persist because we are caught up in the same eugenic script. While biologists have attempted to resolve the nature, mechanism, and significance of variation, fundamental questions of the meaning of variation remain unsolved. The central binary problematic that we began with still remains with us, fraught and contested as much today as then. Nature or nurture? Genes or environment? Science or politics? These questions have never been resolved and have never gone away. Is eugenics feasible biologically? Can we really weed out the weak? Are the poor, the marginal, a testament to weak genetics or social inequality? Indeed, these are familiar frames in contemporary debates. We see a resurgence of biological arguments to explain the privileges of sex, race, class, sexuality, and nation. Neo-Darwinian evolution itself continues to be debated as biologists challenge the gene-centric view. At what level does selection act—base pairs, genes, proteins, cells, individuals, groups, species, communities? Are individual cells themselves multispecies communities (Margulis 1998)? The debates have not changed; only the biological locus of selection shifts. Perhaps this is a sign that science can never refute ideology, that claims of a non-ideological science are delusions.

Yet all these complex histories are entirely erased within disciplinary histories. The architecture of the university relegates questions of nature and nurture, nature and culture, to different sides of the campus. What is apparent is that modern knowledge practices work to conceal, rather than to disclose, the relationship between power and knowledge. A perusal of history shows how internalist histories purge the political and ideological to create a "pure" narrative of the science—the only one that most scientists are taught, if indeed they are taught any history. Hence the ghosts. Indeed, ghosts hate new things.

And yet this history is so instructive. This quick tour of the history of eugenics suggests that, contrary to the mythic archetype of the scientist, the majority of scientists have thoughtfully pondered the relationship of their work to contemporary human societies. They have considered the value of their work and how it may help produce a better society. Perusing this history, the idea that science is devoid of political considerations is laughable. The idea that scientists eschew politics or society also seems plainly wrong. It is heartening to me that we do not need to imagine new worlds—biologists have done so for a long time. We perpetuate the myth of the objective scientist—one who is forgetful, dedicated, and passionate about her or his work—with a willful ignorance and at our own peril. In reality, scientists have a long history of being deeply interested in the world and politics, and have played an important role in social movements and transformations. They have played significant roles and interventions in important cultural and political debates—the holocaust, the bomb, and the Tuskegee experiments. The history of evolutionary biology is filled with scientists quite willing, indeed keen, to deliberate on society. They consciously and thoughtfully worked with the political and social ideologies of eugenics. It was rarely a naive view—these were highly learned individuals, philosophers in spirit, often well traveled (although virtually all western, privileged white men). They came to their ideologies through careful deliberations. The majority of the key players throughout history had strong views and worked hard to enact them. Some of the strongest critics of eugenics were scientists, especially geneticists (Mazumdar 1992, Paul 1998, Roll-Hansen 2010). And yet, these rich histories are entirely erased in biological education.

What is also striking is that eugenics is not a unitary ideology—it spans the ideological right and left. Progressive reformers as well as social conservatives embraced it. In contrast to histories that often characterize scientists as apolitical, evolutionary biology displays a history teeming with the political. Some biologists used eugenics to advocate a diverse and pluralist society and progressive social policy that helped the poor and disenfranchised; these biologists had a vision of a more egalitarian future. Science has at least in the eugenic literature never been far removed from thinking about its consequences for human societies and betterment.

Tracing the history of ideologies of biological determinism is like watching the killing of a hydra. No sooner is one head cut than it sprouts another. The interests of power have always colluded with science to maintain its own interests. Within disciplinary formations, biological determinism has had a strong and consistent current within biology. There appear to be complex intellectual and institutional modes in the histories—from animal behavior to sociobiology, ethology, and most recently evolutionary psychology.

Ultimately, it seems that we can never forget that these are also political battles. There are profound consequences to ideologies of biological determinism and social constructionism. A history of eugenics suggests that both have rather checkered histories. While scientific theories are sometimes embraced, at other times co-opted, certain kinds of politics/power remain remarkably resilient. Ultimately, we cannot expect to fight politics through science—many have tried and failed. But neither can we argue that political battles can be pure themselves. Science holds a central and critical place, and it is the relationship of science and politics that should be at the center of our microscopes; ultimately, both are deeply implicated in power. Neither can escape the other. It is these co-constituted "naturecultural" worlds that we need to elaborate, understand, and work with.

CHAPTER THREE

Singing the Morning Glory Blues

A Fictional Science

> Walking along the great Prospect of our city, I mentally
> erase the elements I have decided not to take into consider-
> ation. I pass a ministry building, whose facade is laden with
> caryatids, columns, balustrades, plinths, brackets, metopes;
> and I feel the need to reduce it to a smooth vertical surface,
> a slab of opaque glass, a partition that defines space
> without imposing itself on one's sight. But even simplified
> like this, the building still oppresses me: I decide to do away
> with it completely; in its place a milky sky rises over the bare
> ground. Similarly, I erase five more ministries, three banks,
> and a couple of skyscraper headquarters of big companies.
> The world is so complicated, tangled, and overloaded that
> to see into it with any clarity you must prune and prune.
>
> —Italo Calvino, *If on a Winter's Night a Traveler*

Prologue

Singing the morning glory blues . . . voice soaring . . . floating through the expansive landscape . . . unhindered, uncontrolled . . . the cadences riding the winds
. . . free floating. No obstacles, not limited by the ground or the sky . . . but free
to explore, to roam the world, all the world. Free to go where the notes take
me, without care, without fears of betrayal or unemployment. Not having to
pledge allegiance to a university, a discipline, or an epistemology. No threats,
no retribution. No accusations of being naive, unscientific, or delusionally
objective. No disdain of playfulness, of stories, of positivism, subjectivism, or
emotionalism. . . . What would it feel like, I wonder. Dare I dream? Dare I try?

How do I imagine a reconstructive project for biology? I do not want to endlessly deconstruct science from the outside, pure and untainted in my epistemological location. After all, scientific work is performed by human actors in relation with their objects and subjects of study. Their rich histories, identities, and social locations, their dreams and nightmares, are entangled in the process of science. I want to keep my hands dirty, tilling the soil to plant morning glories, watching in awe as they grow and then desperately willing them to stop! But I do not want to narrow my vision to this small piece of land. I want to know how this piece of land relates to the city, the country, the world at large, to politics, to poetry and everything else. I do not find the feminist studies of science irrelevant. Rather, they are deeply engaging, productive, and useful. And so, I find myself between disciplines, one that expands my vision, one that narrows it to fine precision.

Inspired by Italo Calvino, I take on the mental exercise of undoing my disciplining. To prune and prune my mental garden . . . to reroute and replace old reference points, those epistemological locations, the familiar methodologies that have ordered my thinking so very easily. Having pruned and weeded and rerouted everything, to learn to live again, to think without those same paths, plants, ground, and walls. The exercise is excruciatingly difficult—I am haunted by memories of my disciplining. Whoever thought it would be so much more difficult to unlearn than learn? I persist.

And so, gentle reader, kick your shoes off, put your feet up, find a nice warm cup of tea or better still a glass of wine and join me in this story told in three parts . . .

A Reader's Guide

As I begin a reconstructive project, such as this, you might well ask: what form, style, and analyses can she use that draws from both the sciences and the humanities but still does not belong to either? My answer here is fiction. The story you are about to read traces the intellectual adventures of three high school girls who encounter a group of researchers. The researchers have just begun work on a morning glory field in the girls' town. Each of the researchers asks particular questions that are disciplinarily based. Each is convinced that his or her question is the most important and the most methodologically sound.

The story is told in three parts. *Part One, The Encounter,* introduces the researchers and their individual perspectives and research questions. In *Part Two, The Debate,* the researchers are forced into a conversation with each other. What might the disciplines have to say to each other? Can they learn from each other to enhance and expand their disciplinary perspectives? In *Part Three, The Synthesis*, I present a fantasy where I locate the three girls in a futureworld of an alternative science, a fictional science. I believe we need not only science fiction, but fictional sciences—imagining other configurations of knowledge making, reconstructing alternate inter- and a-disciplinary lenses, new conceptual practices, and more engaging plots and stories that are located in the interdisciplinary fissures of the sciences and the humanities. As Faye Harrison (1996: 234) argues, "Fiction encodes truth claims—and alternative modes of theorizing—in a rhetoric of imagination." Now, on to the story . . .

Part One: The Encounter

I don't like jargon. In fact, the more I write and the older I become, the more I abandon it, by a progressive effort toward the greatest possible clarity. Technical vocabulary seems even immoral: it prevents the majority from participating in the conversation . . . You can almost always find a lucid way to express delicate or transcendent things. If not, try using a story!

—Michel Serres, *Conversations on Science, Culture, and Time*

The story begins in a small town on the outskirts of Chennai called Thirumbaram. And our heroines are Tara, Tulasi. and Tabrez—the Thirumbaram Three, or T3, as they were affectionately called. The three, friends since childhood virtually grew up together and were jokingly called "the national integration project." Like state propaganda encouraging India's diverse religions, castes, and linguistic groups to get along, the three personified this mission since they came from different and often mutually hostile religious, linguistic, and caste communities. Into their deep, abiding friendship they brought cohesion and stimulated interactions and together created a strong and diverse community. It was a warm summer morning. Dawn was just breaking. T3 strolled down a dusty side road, savoring the freedom of their summer holidays. Their *pavadais* swished around them, their pigtails swinging in the gentle morning breeze. They approached their favorite tree, which had three perfect perches. It had taken them many a summer to find this favored spot; hours of pruning gave each a perfect view of the stunning field of morning glory flowers. The morning air was crisp and the gentle breeze created an illusion of a morning glory field dancing in a beautiful hue of blue. This was their favorite part of the day, staring down on the sea of flowers, glorious in the morning sun. A few hours from now the sun would blaze down on the field, flowers wilting to a quick one-day death. New flowers would emerge the following day, just as spectacular. It was the metaphor of the dying flowers and their dwindling holidays that captured their fascination. They spent much of their holidays outside their homes. Their curious minds experimented with plants and unusual physical phenomena that captured their imaginations. They were in fact quite notorious in town for their exploits. But their charm always won people over. People often called on them to fix their household gadgets, help ailing plants, or repair bicycles or cars. Their reputation grew each year.

This morning seemed like any other morning. They nimbly climbed up their favorite tree, only to find that they were not alone. There on Tara's favorite perch sat a man who was as startled as they. They quickly glanced down to realize that the field looked different. Several people were crouched down in rapt concentration. Curiouser and curiouser, they thought. They turned their attention to the matter at hand, to the man on their favorite perch.

The man was sitting with his legs hanging down, holding a book into which he was scribbling furiously. His skin was as white as snow, or the way they imagined snow at least. He was wearing clothes with blotches of green and brown. Camouflage gear, our heroines thought. He was hard to spot. The man seemed startled by the three menacing faces demanding to know who he was.

"I am the Dhanush of Manush," he said. "I am studying all those researchers down there. Observing what they do, watching how they study what they study, whom they talk to. See my data collection book? I score their activities. They study plants, bugs, soil, and rocks, and I study them. You see, we are an international group funded by the Bored Foundation of the U.S.A. What is really unique about this project is that it tries to bring researchers from different fields to work together. Usually, each of us works alone and then publishes our work in our own disciplinary journal. But this project allows us to share our insights and perspectives with each other to create something new. Might the whole be more than the sum of its parts? This is the first of the experiments," he said importantly.

"So what are they studying?" asked Tabrez.

"Why don't you go ask them yourself?" he said, bursting into a fit of nervous laughter. His face scrunched up and his eyes peered brightly through it. It was an unnerving sight. Not wanting to show their discomfort, they chorused, "We'll be back," and deftly scaled down the tree's trunk.

• • •

They approached the first of the researchers, The Vidhvan of Vigyan. He was skillfully inserting a small tube into a flower. His clothes were muddy, his jeans frayed at the knees. His skin was tanned and tough looking. He looked up as the girls approached.

"What are you doing?" asked Tulasi.

"Well," he said, "I am extracting the nectar from these flowers. I will then take this fluid to the lab and measure the sugar content. Why? You see, I am working on morning glory flower color variation. See the variation in flower color? This is one of six sites we have chosen to work on around Chennai. We want to understand why this variation in flower color persists. Morning glories are hermaphroditic—that is, each of the flowers has male parts and female parts."

"Really, male *and* female?" Tabrez countered.

"Well, to be accurate, I should call them pollen and seed vehicles. But that gets tiresome sometimes and difficult to explain to little girls like you." He smiled. "You see the field in front of you? The flower colors exist in a particular ratio. I am interested in whether these ratios are in equilibrium—do they stay the same frequency year after year, and will they return to the same frequency if perturbed? What makes my work so unique is that I don't just count seeds or the female contribution to fitness. I also look at the male contribution, that is, pollen. Bees are central to this. I watch the patterns of bee visitation, which flowers do they visit and in what order. Variation in sugar content of the nectar can be important in bee choice. When bees visit a flower, they transfer pollen from that flower to subsequent flowers they visit. So, if they visit a blue flower first and then a white flower, they are taking alleles for blue flower color to white flowers."

"Why can't you just come back and watch this field next year and the year after?" asked Tabrez.

"We'd like to. But then I can't publish papers for years. I'll never make tenure!" he grimaced. He continued, "Who's to say the Bored Foundation will still want to fund my work next year or the year after?"

"You mean you are trying to understand long-term evolutionary factors but study it just in one year?" asked Tara.

"Well, it's unfortunate. But you can try and get a snapshot of evolution. I wish I were rich and free to squander my wealth away, with a large house, a huge field in my back yard . . . ," he went on dreamily only to realize that the girls were gone.

• • •

The Raja of Baja was lying on the ground staring at the dancing flowers, his eyes transfixed on the field. His white *veshti* was tied loosely around him, his *thundu* curled around to serve as a pillow under his head. The whiteness of his clothes stood out starkly against his dark complexion. It was not an unfamiliar sight in these parts. He sat up as the girls approached. They inquired after his work. "I'm one of the resident poets and musicians," he said. "I am to be inspired by these beautiful flowers, the wondrous scenery, and translate these into words and song. This is the life I always dreamed of. Who usually gives you money to do this? I'm enjoying every minute of it." He closed his eyes and started reciting his poetry. The words flowed in a gentle rhythm, capturing the singsong Tamil language. The girls listened in rapture. The music was most haunting. His voice rose, the breeze carrying the lilt of his voice across the field. A man they had not met yet rose from the field a few yards away and told him to be silent.

The Raja shook his head. "They have no soul," he barked at no one in particular. "Stick their heads in the mud all day, coming up with pages of numbers. Look around you. Who can reduce this beauty to such numerical nonsense? How can you sit in this field and not be moved by this glory?"

"But maybe it's a different kind of beauty they see," suggested Tulasi. "You both seem to draw your work from the same field, don't you? Maybe you tell different kinds of stories."

His face suddenly grew animated. "I write about beauty, the soul, and the greater meanings of life."

"Do you ever write about evolutionary change, how rocks weather to soil, how transposable elements move around in the genome? Seems like it's wonderful material," said Tara.

"Oh, no! Scientific language is unpoetic, and honestly, science was never my strong point. I never did well in the sciences in school," he said sheepishly. "But scientists have all the power. When they work, it is pure and true. Me? I'm only a storyteller," he said bitterly.

· · ·

They walked toward the man who had yelled at the Raja. The Kahani of Rasayni was an immense man with a commanding presence. Around him were three students taking samples of the soil. He seemed to supervise their activities carefully. "My graduate students," he explained. The students handed him the samples and he put them away in specimen containers. "I'm sorry I yelled," he said, "but the damn fellow was breaking my concentration."

"What's in all these tubes?" asked Tabrez.

"I'm analyzing the soil in this field," he said. "This is a very interesting area geologically. And I have become fascinated by its soil. I'm interested in measuring the heterogeneity in this field. In this area you see bands of soil types. This area is very recent, maybe two thousand years old. A most intriguing finding. But the geological record is not very well preserved. Much of it was destroyed. We would love to find a better history of climate conditions here. But the written scientific record is poor."

"How about reading the literature of the time? Exploring mythologies, listening to the oral histories of this region? It's very rich you know," said Tabrez.

"That's the work for historians. We work with rocks, only hard data," he said dismissively.

"So you won't enjoy the view like the others," said Tara, changing the subject. "You'll be stuck in your lab."

"Come the afternoon heat, we'll see who is complaining!" he said. "I like lab work. It is precise, exact. You get the right value and you can replicate it any number of times."

"But," said Tara, "the Raja was saying that it was just a different story than the one he tells."

The Kahani's face turned red. "Nonsense!" he said. "How can you compare that to what I do? If you measure the nitrogen in the soil, it's exact. That is science. But to compare that with beautiful words and elegant phrases? I appreciate a good poem, a good story, before I go to bed. That entertains, relaxes. But . . . ," he was shaking his hands furiously, "that is not science. We do not tell stories. I don't make up numbers. This is fact, clear, irrefutable data. Get it?" The Kahani was becoming increasingly agitated, now towering over them, his finger jabbing the air in front of their faces. The three girls nodded and left quickly.

• • •

The Pundit of Poojyam[1] was admiring a flower as they approached him. "I heard your conversation with the Kahani." He smiled. "What emotionalism. What he does not realize is that we are all part of the same act of creation, all a particular finite number. The world is best described by mathematics. Take the number zero. It epitomizes the absolute, the all encompassing, or in our Hindu philosophy, 'Nirguna Brahman.' This absolute can never be comprehended by words, can never be described by us mortal humans, and is beyond the comprehension of the human mind. Our small, inconsequential mind. Take 'infinity,' the cornucopia of reality, which has all the possibilities that we can see in our reality. Now when you multiply zero and infinity, you get the entire supply of finite numbers . . . and you and I and everyone else is the product of one such union."

"So, we humans and mathematics and science are all part of a divine order?" asked Tulasi.

"Yes, algebra and geometry and the rest of mathematics constitute a language, a grammar by which to write the story of the cosmos. You know," he said, "I sometimes have dreams where the god Narasimha talks to me through scrolls covered by the most beautiful theorems. My life and my work merely describe the grand design of life."

"What do other people think of your work?" asked Tulasi.

"Well, my mathematical work is well received. I am considered to be quite brilliant. But no one recognizes what this work means to me. They brush it aside, unwilling to listen. All they care about is abstract numbers. Not what those numbers mean, their divinity, their cosmic connections . . ." His eyes suddenly glazed over as he stared into the sky, lost in thought.

• • •

The Sheikh of Leikh looked wistful as the girls approached her. Another researcher, scribbling away! They use up more paper than the entire town of Thirumbaram, Tara thought. The girls approached her. The Sheikh was the linguist of the team. She explained, "I am interested in language—words, metaphors, and analogies we use to describe the world around us. What we know, and see, is always mediated by language. Our oral and written traditions are all mediated through words."

"What does it matter what words we use?" asked Tabrez. "Aren't we describing the same thing? As the Kahani of Rasayni argued, whether we called it an insect or an oolong, isn't it the same creature?"

"Well, not quite," said the Sheikh. "It may be true that the insect remains the same creature with its mosaic eyes, pair of antenna, six pairs of legs, et cetera, but we give meaning and symbols to creatures as we name them, and these words take on particular significance and meanings. For example, we belong to the class Mammalia. Mammals are named mammals because when Linnaeus was naming his classification system, there was a big campaign in England to promote breastfeeding. Therefore, even though mammals are defined by many characteristics such as hair, a four-chambered heart, a single-boned lower jaw, three middle ear bones, a diaphragm, and mammary glands, and although they all maintain a high body temperature, it is the feeding of the young that came to define us. I would add that it is not accidental in a culture where we name nature as female, that the focus on breastfeeding would only add to our self-definition as a species where women are the caregivers. Where women are taught to sit at home and take care of babies and cook and clean. Words are powerful; they tell us how to think of ourselves. If you ask kids what a scientist looks like, they inevitably draw a white man with a beard, glasses, unkempt hair, and a white lab coat, even when female figures in their families might be scientists! There is nothing innate in what we know, it is all learned."

"So you think that scientists are oblivious to both the consequence of their language or how their work is a consequence of language?" Tara asked.

The Sheikh smiled. "That's beautifully put. Will you go explain that to them?" she said as she burst out laughing. "They never get it!"

• • •

The Kondai of Mandai had just finished interviewing the Mantri of Tantri. Like her name suggested, a humungous *kondai* sat on her head. This sight transfixed the girls, who wondered when her head might give way to the weight of the massive ball of hair above.

"Well, I am the psychologist of the group. I am studying the individual psychological motivation of these researchers in doing the work they do. What propels them to leave their hometowns, or in some cases cross the seas to this godforsaken place?" T3 were not amused with their town's description. The Kondai went on quite obliviously. "I cannot of course go into the details of the personal lives of all the people here, but when I interview them, each of them lets escape their motivations. Nothing can miss my eye," she said shaking her finger.

"But what about random chance and serendipity?" Tabrez asked.

"Individual narratives are notoriously inaccurate. People constantly reconstruct the past. My conclusions come from very strong evidence in my field. Robust findings. Whether these individuals recognize my conclusions is irrelevant. I am right."

• • •

Skipping down the field, they almost collided with the Saraswati of Sansthithi. She was crouched low, examining the amount of damage on a morning glory leaf by a herbivorous insect. "I am testing whether and how natural selection is acting on this flower color variation," she said.

"How do you measure that?" asked Tara curiously.

"Oh! Let me explain." She sat down on the ground and the girls sat around her. "Charles Darwin, the founder of evolutionary biology, made two important theses in his famous book, *Origin of Species*— descent with modification and the process of natural selection."

"Huh," said Tabrez.

"Think of it as a slower version of artificial breeding. In fact, Darwin did a lot of observations watching artificial selection of pigeon fanciers. How do you get such a variety of pigeons, of dogs, of cats, of cows? Breeders do this by selecting for particular traits and then making sure that individuals with those characteristics reproduce with each other. It is amazing how quickly we can produce distinctly different animals and plants."

"So who does the selection in nature?" asked Tulasi.

"Well, Darwin argued that it wasn't about any creative force but about individual variation, adaptation, and selection. So, for example, in this field, all the plants will produce seeds and die. Next year, those seeds will grow. One plant might produce no seeds at all and will leave no offspring, another might leave twenty offspring, and yet another twenty thousand. Therefore while the ratio of the three plants is 1:1:1 in this generation, they will be 0:20:20,000 next year. While Darwin did not know about genes or Mendelian inheritance, we now understand how, for example, purely by chance, you may get a field of all blue or white flowers," she said, her face red with excitement.

"But what if bees liked blue flowers and not white flowers? Or if white flowers produced more seeds?" asked Tara.

"Aha!" said the Saraswati. "Then natural selection will favor and select the blue in the first case and the plants with white flowers in the second. But it won't be about chance. Or there may be selective advantages to both and we might find an equilibrium of both colors to sustain a flower color polymorphism, as we have here."

"Do you think that is the case here?" asked Tabrez.

"That is what my intuition tells me. Only through developing carefully controlled experiments can we reject some hypotheses."

"How important is intuition in science? We were talking to the Kondai of Mandai and she was suggesting that it would help scientists to reflect on the origins of their beliefs," said Tara.

The Saraswati smiled. "Forgive me, but that's psychological babble. The beauty of science is that it doesn't matter. Ultimately, you need produce experimental proof. Period. Psychology has no contribution to make to science. None whatsoever."

• • •

It was getting close to lunchtime. And the girls spotted a man with a big white hat approaching the field. The Surdas of Ithihas was accompanied by a man who carried a large bag filled with lunch. The girls rushed to greet him.

"Why aren't you here studying the field like the rest of them?" asked Tabrez.

"I am the project historian," he said. "My work keeps me at the archives of the local museum."

"You mean if someone had discovered something different one hundred years ago, you would all be doing very different research today?" asked Tara.

"Exactly. But not everyone agrees. For example," he continued, "some of the scientists here would say that even if Darwin had not postulated his theory, someone else, like Wallace, eventually would have because the forces of natural selection are real. This is the overdeterminationist position. I, however, take an underdeterminationist position—that history takes a particular direction only because of particular events. A different set of events? Different futures."

"But," said Tulasi, "wouldn't some questions be obvious? Wouldn't someone eventually have noticed the falling apple or discovered that the earth was not flat?"

"I'm not denying that some discoveries are inevitable. But depending on the cultural moment, their meaning and significance shifts, as do the frameworks, theories, and cultural beliefs. Science is not a linear progression to greater truth. You see, new ways of looking at the world can completely change and transform everything we know. There is a relationship between knowledge and power. Louis Pasteur's experiments alone did not account for how profoundly he changed seventeenth-century France. His success depended on a whole network of social forces. We need to give up the model of the lone scientist battling the world on his own, and the belief that truth ultimately conquers all. The world does not and never did work that way."

He continued. "You see I work on the history of the 'gulisthan particles.' They are tiny particles that were discovered in cross sections of rocks about one hundred years ago. This discovery transformed the subfield of gladiolar geology. Because these particles change the composition of the rock around them. It was unheard of. But what does this say about the field of gladiolar geology before this discovery? Did gulisthan particles exist? Did scientists just miss them? Or did their slide preparation techniques not allow them to see these particles? What was it about that period that allowed for such a discovery? It is fascinating when you locate scientific discoveries in their social context."

"But" suggested Tara, "couldn't you just take the slides from the periods before to see if you could spot the gulisthan particles even though scientists one hundred years ago could not?"

"Ah! But you see I'm not scientifically trained. I don't know how to do that. And to scientists it is irrelevant."

He quickly glanced under the tree where all the others gathered, waiting hungrily. "I really ought to go," he said quickly.

• • •

The Mali of Pali was taking a break. He was lying flat on his back, his eyes closed. A bottle of sunscreen stood on the ground beside him. He woke up upon hearing the approaching footsteps of the three girls.

"Hello," he said. He began to talk about evolution and Darwin when the girls cut him off, telling him they knew all about Darwin.

"Ah! You've been talking to the great Saraswati, I see," he said.

"Do you do the same work that she does?"

"Well, yes, but our intuitions are very different. What she says is important. Equilibrium models are important in ecological systems from a global perspective. But scale matters. So if you look at islands, ecologists would argue that the number of species on an island remains relatively constant while the identity of the species present changes continually. Yet here, I'm convinced that equilibrium models of evolutionary change are of practically no use. For equilibrium models to work, you need some sort of stability, where the same selective forces can act year after year. Look at the weather record of this region. It's volatile—plagued with droughts, floods, hurricanes."

"Yeah," the girls said. "They are predicting heavy rains again next week."

"Nature seems much too random, much too unpredictable here. It is a pity that the meteorological records are so poor in this area," he sighed. "Nothing to be done about it."

"You know," said Tulasi, remembering, "there is an old man here called Ramabadran. He has been keeping records of the rains and the temperature for the past eighty years. Was a hobby of his from childhood."

The Mali grew thoughtful. "It is unlikely that we can use his data. Data have to be calibrated accurately to be useful. No scientific paper would accept such observations. Could I continue my midday nap? I am exhausted."

• • •

The Mantri of Tantri pulled down his spectacles as he saw the three girls approach. He sat on a fallen tree. "I am a philosopher," he said, "and I am interested in how we know what we know."

"What do you mean?" asked Tara. "Could you give us an example?"

"Certainly," he said. "Take the claims of science that it produces real, value-free, objective knowledge. That it describes nature precisely and exactly. History proves otherwise. We now know it is deeply entangled in politics. Take the whole history of eugenics, or the scandal of Cyril Burt and the IQ tests. Science is not magically exempt."

"So who controls science these days?" asked Tabrez.

"It's not an easy answer. Some would say the corporate world and defense-related research. It is unfortunate what has happened in the name of development in this country. With colonialism came western science and technology and we the colonized have taken to it and embraced it whole-heartedly, without critique. We have tried to be better *sahibs* than the English. We have delegitimized the rich traditions of indigenous science and medicine even while they enjoying a boom in the west. It is ironic, this postcolonial condition!" he said, shaking his head.

. . .

The sun was setting and the girls had had their fill of information. It seemed so much to take in during a single day. A woman sat at the edge of the field watching them. Her sari was wrapped around her head to shade out the sun. "I am the Rani of Pani," she said. "This land used to belong to my family."

"What happened" asked Tulasi.

"We were farmers. We put a lot of our money into buying new technology, high-yield seeds, fertilizers, pesticides, and herbicides the same year as a big drought. The following year we were greeted with a flood. My family never recovered from that. Here I am without land, without food, and there are all those people studying a field of weeds," she spat out. "Weeds! Can you believe that? What a waste of land."

Part Two: The Debate

> You seem to think that no idea exists or blooms except in opposition to another or others . . . An idea opposed to another idea is always the same idea, albeit affected by the negative sign. The more you oppose one another, the more you remain in the same framework of thought.
>
> New ideas come from the desert, from hermits, from solitary beings, from those who live in retreat and are not plunged into the sound and fury of repetitive discussion. The latter always makes too much noise to enable one to think easily. All the money that is scandalously wasted nowadays on colloquia should be spent on building retreat houses, with vows of reserve and silence.
>
> —Michel Serres, *Conversations on Science, Culture, and Time*

Several days elapsed. The girls developed a routine of visiting the field each day, watching the researchers and growing fond of them. But researchers usually talked to them individually, never together. On the rare occasion they did, the researchers inevitably grew frustrated with each other. The researchers continued to work in their isolated and individual worlds; some divisions seemed sacrosanct. The girls found this most curious. Having eaten lunch with the researchers each day, they invited the entire team to dinner at Tulasi's parents' restaurant. Quite a feast! A spread of *idlis,* masala *dosas, sambar,* various chutneys, *puliodarai, bisibele bhat, vangi bhat, bagla bhath,* sumptuous *payasam* . . . the spread went on. The spicy aroma filled the room, the colors mesmerizing.

Everyone dug in, appetites piqued and gradually happy, contented bodies lay on the floor, satiated. It seemed the ideal moment for the girls to get all the researchers talking.

"When we were walking back the other day," began Tabrez, "we met the Rani of Pani. What do you think of her story? How do you rationalize your work in her presence?"

"Well," said the Mantri of Tantri, "it is most unfortunate. But it is what western science and technology has done to this country."

"Come on," said the Mali of Pali. "You surely cannot write off all of western science and technology with that. Do you realize that the farmer who owns the field we use makes much more money renting it out for cultivation? I have little patience for those who blame all the ills of India's government and its stupid and corrupt policies on western science and technology. What was so wonderful about the India of the past that we want to hold so sacred? Isn't it mainly the rich activists who make such arguments? Why cannot the poor enjoy the benefits of technological growth?"

"You misunderstand me," said the Mantri of Tantri. "When I say western science and technology, I include the Indian government. The invasion of western science and technology is a colonial legacy, as is the corruption and bureaucracy in the government. How can you argue with that?"

"Sure, it's a colonial legacy," agreed the Mali of Pali, "but is it about western science or how we use it? Can't we use science to benefit India? Aren't machines more humane than humans doing backbreaking work in hundred-degree weather? Science and technology have really improved the quality of life everywhere. We need basic research. While morning glories are a weed species, we are uncovering fundamental processes that govern nature. How will we learn to conserve our forests? How do we feed our overpopulated world?"

"You are missing the point," said the Dhanush of Manush. "I don't think the Mantri is against basic research. It's about the power of science and technology. Scientists have a carte blanche to do whatever they please. We have more than enough food to feed the world today, yet people go hungry. And don't the humanists improve the quality of life also?"

"We don't expect to pass judgment on your poems and literature," said an agitated Mali of Pali. "Stay the hell away from our experiments. You know nothing about it." With that, he left the room, disgusted.

"I think we are confusing issues here," said the Saraswati of Sansthithi, trying to sound reasonable. "I think we are confusing the content of science from the context of science. Now, I completely agree that the context of science needs changing. That we need to examine the directions we take. Whether we ought to continue with nuclear power, how we spend public money. I also agree with the Dhanush of Manush when he points out that the culture of science is fraught with problems. As a woman working in the so-called third world, I am deeply aware of it. I am not respected, easily dismissed as a woman. And in international conferences doubly dismissed because I am from the third world. We have limited resources and poor networks. I cannot afford to work on questions the west considers the cutting edge. I find research problems that are less flashy but interesting. Listen to all the names of the famous researchers we quote. They are all from the west and largely male. So, I relate to arguments about needing to change the context of science. But," she continued raising her finger, "this does not mean that when I undertake a problem, I approach or solve it any differently. It is a universal science, a universal scientific method. When you start making claims of scientific content and interpretation, I disagree there."

"Come, come, now," said the Sheikh of Lekh. "That is much too general a claim. Take, for example, the work of the Vidhvan of Vigyan on the 'male' contribution to a plant's success. Why male? Why take heterosexual norms of human culture and impose it on plants? Don't you see how we anthropomorphize? How the pollen grains now start fighting tooth and nail as they penetrate the stigmatic surface and make their way down the pollen tube? They are metaphors about brave male warriors, not descriptors of the plant. And scientists claim they are being objective? Really!"

"But we never claim to be infallible," protested the Saraswati, "nor to be producing the absolute truth. No scientist worth anything makes such claims. We try to be objective" she said stressing, "try," "and most of us realize that we don't always succeed. But what we always strive to do is to produce the best possible knowledge we can. Incorrect information, fraud, anthropomorphisms—they get discovered eventually. Science self-corrects. That is the power in it."

"Don't you realize," cut in the Sheikh, "that self-correction is contextual itself—it only seems that way because it conforms to a new cultural and political order?"

"I don't think that is always true," challenged Saraswati. "But exactly how do you prove your theories? How can you tell that scientific progress would have been any different with a different language or discovery?"

"No, we can't," agreed the Surdas of Ithihas. "But there are many ways in which to produce knowledge. True, it's an unfalsifiable hypothesis. So what? Can't we draw inferences on the meaning of scientific discoveries? About language? The relationship of knowledge to power?" The Saraswati shook her head and left.

"Getting back to your point," said the Vidhvan of Vigyan. "I will grant you that we need to be more precise with language. But what's in a name? The example you give is blatantly bad science. You ought not to be anthropomorphic. Any scientist will agree with that. But I still don't see what difference that makes."

"Okay," said the Sheikh, "take your example. Is it accidental that so far we have only focused on the female contribution? Isn't that because we see females as caregivers? Talk about reproduction and we think 'female.'"

"Absolutely not!" yelled the Vidhvan of Vigyan. "We have not had the technology to follow pollen grains. It's only with the advent of molecular biology that we can create molecular markers to do this work. Do you know how much more difficult and laborious this is? Do you know how tiny pollen grains are? Some things are just not possible!" he said, walking away followed by the Sheikh.

"I am not entirely convinced by that argument," Raja of Baja interjected. "Things seem impossible to you because you are caught up in your own framework. Technology is not the only thing that can solve problems, sometimes it takes a shift of mind."

"Clearly you know nothing about science," retorted the Kahani of Rasayni. "You keep accusing science of being too dominant and yet you give to it powers that it does not and cannot possess."

"We are not magicians," added the Mukhiya of Sankhya. "Before the advent of computers, we could not solve certain problems. They were too complex. It was just not possible. To say that we should have sat down and spent ten years solving one problem is just plain silly!"

The Saraswati and the Kalainyar walked out together, the Dhanush and the Mantri closely behind, eyes averted.

"You know," said the Mukhiya of Sankhya, " I find this refusal to accept scientific facts and laws most curious. Why are you so resistant?"

"Well," said the Kondai of Mandai, "of course we accept some scientific facts and laws. Of course I would not jump off a building and expect to fly. But it is what we embed these observations in that interest us. We shrine them in laws, sometimes being careful to specify the conditions under which they operate. It's the embedding, the need to universalize everything, the impulse to control and manipulate at the stroke of the hand." The Mukhiya walked out.

"I have no idea what you just said," said the Pundit of Poojyam.

"You expect me to labor through all your statistics," protested the Kondai, "but you won't learn my language? Humanists should always be immediately transparent! We can't have jargon, our own vocabulary, theories, methodologies? This is impossible! The same issues, again and again," she muttered, leaving.

"I want to bring up the issue of aesthetics and beauty," said the Raja of Baja. "To me, as a poet, they are paramount. Each word is deliberate, and each sentence carefully crafted. Aesthetics is above all important. But you scientists reduce it all to numbers."

"Wrong again," exclaimed the Pundit. "What's the use?" He stormed out.

"What he meant," said the Surdas of Ithihas, "is that beauty and elegance are very important in science. It is common to talk about an elegant experiment, an exquisite theory. Simplicity is extolled. Scientists go into rapture in the presence of elegance. Read the memoirs of scientists. They are passionate, immensely creative beings. You know, that among the famous geneticists, Dobzhansky compared natural selection to a composer, Ernst Mayr to a sculptor, and Julian Huxley to Shakespeare! There can be something in common between us after all."

The Surdas of Ithihas and the Raja of Baja left together. The girls grinned. What a melodramatic evening! The emotional outbursts, the wounded egos, the grand stances. Stifling their giggles, they helped clean up and went home.

• • •

Several weeks elapsed since the fateful evening. The egos calmed, and the cold shoulders warmed. Yet, the air of hostility persisted. The researchers had agreed to disagree. At the end of the summer, they left. The T3 received the published work. There were few collaborative papers. In this sense, the project, they heard, was considered a failure. The summer ended and the girls went back to school. The years sped by. This extraordinary summer, however, unknown to them, was the beginning of something wonderful.

Part Three: The Synthesis

How do you live and think together beneath a light that warms our bodies and models our ideas, but which remains indifferent to their existence? We contemporary philosophers cannot ask this question while ignoring the sciences, which, in their very separation, converge to ask it, even to exacerbate its terms.

And when 'the world' means purely and simply the planet Earth, . . . when humanity is finally solidary and global in its political existence and in the exercise of science, it discovers that it inhabits a global Earth that is the concern of our global science, global technology, and our global and local behaviors. This is the reason for the necessary synthesis.

—Michel Serres, *Conversations on Science, Culture, and Time*

We return to the same morning glory field three decades later. We arrive at the Chennai airport. Buses and trains fueled by solar energy await eager passengers. The town seems to have lost its name to the famous Saraswati Institute, now a decade old. Thirumbaram is transformed. It is bustling, with an air of activity and excitement. The world-renowned institute brings scores of researchers across the globe. We hear that the institute was conceptualized and constructed by the Thirumbaram Three as a unique project of synthesis.

As we probe, we learn that two decades ago the Thirumbaram Three were granted the first joint, group Ph.D. in India. It was a unique event. While they had each specialized, they were broadly trained as generalists. They were known to be the founders of a new movement of synthesis. Saraswati, the Goddess of Knowledge, provided that larger rubric, the impulse to create knowledge, unbounded and unhindered by disciplinary barriers. As we enter the Saraswati Institute we are impressed by the functional elegance of the building and its decor.

The space houses a large number of people, yet allows for private conversations and discussions. We see schoolchildren working on projects, touring the facilities. Researchers bustle along. The center of this multi-winged structure is a huge hall. The dome-shaped ceiling envelops you the moment you walk in. There is something immediately warm, energizing, and inviting about the room. Maybe it is the lighting, or the architecture? We are not quite sure. We hear that this room has been nicknamed the "well of answers" because so many problems have been solved here.

We step outside to the field. While it was originally one small field, the area has grown. Our eyes feast on acres of plants dancing in the morning breeze. It is quite a spectacular sight. We notice that there are plots that house other plants. It is not clear exactly what those are. As we walk to the field we see the T3 in discussion. They are delighted to see us and eager to talk.

"Well," began Tabrez, "we'll try and summarize quickly what we've been up to for the past thirty years. My! How time flies!" she says, laughing. "Our approach to research has been marked by collaboration, both among ourselves and with the town itself. We were not interested in creating an institute as an ivory tower, removed from the lives of the people and with little meaning. We wished to involve them in the work. We grew up in this town that has been very kind and generous to us over the years."

"You can call it a kind of symbiosis, if you like," said Tara. "We need them and they need us. We have brought a great deal of respect and fame to this town. Our work is driven by world problems."

"We always wonder," said Tulasi, "why our results have been so spectacular. Is it our long-term collaboration? Open minds? Communal training? Broad and interdisciplinary training? Collaboration with the community?"

"You see, after the researchers left that summer, we began studying the morning glory field ourselves," said Tabrez. "Tara got fascinated by the literature that came from this region. They were written in this particular dialect that Tabrez's family speaks, and they were invaluable in helping her translate. Tabrez's long interest in literature was invaluable." They all smiled in nostalgia. "Along with these stories, we also reviewed town meeting transcripts. We talked to town seniors. What fascinated us were repeated descriptions that were hauntingly similar—visions of green and yellow swarms. We saw it appear every twenty years or so, rather consistently. It was most curious."

"Tulasi is a big fan of this man called Ramabadran. Her family and he had a long, established history. Ramabadran's community followed animist traditions and have deep connections to nature," continued Tara. "Ramabadran kept meticulous records of the weather and gave us access to them. Among other data, this man who was in his nineties had been collecting records of rain, temperature, humidity, and other weather patterns since he was a child. Tulasi, you know, has quite a mathematical bent. I don't know how she sees these relationships between numbers. Quite escapes me. But anyways, she realized that with the right coefficients, a combination of the weather variables created a thirty-three-year cycle. In this cycle, the monsoon winds were crucial."

"Then one day it happened. Right before the monsoons, for the first time in our lives, we witnessed *Morisuca frambia* swarm into town. They were green-colored lepidopterans. We had heard about this, and it was well known, but no one had noticed the pattern of their visitation. And of course, no one noticed that they laid their eggs on the blue morning glory plants and that the larvae then fed on these leaves, decimating them. Tara did the necessary chemical analyses of the cyanidins, showing that compounds in the blue pigmentation pathways were crucial for the *Morisuca frambia*'s survival. Without that, the adults are infertile. As a result, for a few generations, the blue-colored morning glories do very poorly and the white ones take over. Now the feces of the *Morisuca frambia* have an unusual compound we called *frembia*. Growing morning glories with this compound particularly affects the white genotypes. We showed how these effects, even though environmental, are indeed passed down epigenetically to future generations of white genotypes. It was this non-Darwinian mode of inheritance, where an environmental character fundamentally altered the expression of genes for many generations, which aided the proliferation of the white genotypes."

"And the wonder of it is that a few decades later, you get a yellow fungus, *Gorima stabia*, which thrives on a compound in white-colored flowers. And here, Tara's background in virology proved invaluable. Tara comes from a family of ayurvedic doctors who have a unique specialization in diseases, especially a local illness called *Gorima pani*. *Gorima stabia* carries a virus that causes this local fever. The virus affects the mitochondrial DNA of white-colored flowers, inserting elements into it. For a few generations following the yellow swarms, the white genotypes find their fitness severely depressed. Again, a complex non-Mendelian mechanism through the mitochondrial DNA, which in concert with the nuclear DNA depresses growth of white genotypes. And so the pattern goes. First, the blue flowers affected and then the white. Now, you are probably wondering what causes these cyclical patterns." Indeed, we are.

"Tabrez, our weather whiz, figured that out," said Tulasi

"Well, the trade winds are really crucial to this. Once the lepidopterans have decimated the morning glory population, the trade winds carry the adults to the next spot. We actually followed the path of the winds to show similar patterns in morning glory fields right across its path. And indeed within a few decades, they return to this same site and by then there is enough food for them again and so they cycle. Amazing! If not for weather data of Ramabadran, the historical and literary data, and indigenous medical knowledge, we would never have stumbled on to this. It created quite a stir when we figured out the whole story. The stochastic pattern of the weather creates variations on the themes—only in pulling together local knowledge of weather patterns, lepidopteran history through Tamil literature, and the well-documented indigenous knowledge practitioners of Ayurveda could we figure this out. It was quite something."

"And then, of course, came Tulasi's great theory that also created quite a reputation. Tell them, Tulasi."

"Well, we were doing some cage experiments to test our hypotheses. We were doing our master's degree then. And Tara had just read Foucault's *The History of Sexuality* and we had had a really intense discussion of it that night. And it really got me thinking about the ways in which we measure a plant's reproductive success or, namely, its fitness. And so we had to follow both the contribution of the pollen parent and the seed parent to really get an accurate estimate. Tara, bully that she is, got me thinking about the ways in which we introduce heterosexism even into plants. I mean, here are these creatures that produce pollen and seeds that we immediately assign into a binary male/female and trace their lineages just as we do in human beings. Should we really read the same kind of sexuality into plants? Here are hermaphroditic creatures—that is so cool! I began wondering if there wasn't another way to measure fitness.

And so one day I was running these gels to figure out the pollen parents and made a serendipitous but glorious mistake in using a wrong chemical and got the most amazing pattern. This critique of sexuality hot in my head, I investigated further. To cut a long story short, I discovered a set of markers that were unique in plants, sort of like fingerprints that can be inherited. And so you can trace lineages without having to think of binary male/female or even pollen/seed. And this was so freeing. You could track horizontal and vertical transfer of genetic material. You could now construct models where passing down genetic material was a communal dance. How perniciously promiscuous plants were—not only with their heterosexual counterparts, but a host of other microorganisms. How much information we lost in seeing plant generations as heterosexual reproduction! And so I developed a new theory and mathematical models to go with it," said Tulasi modestly.

"Indeed, Tara's family background was important to challenging traditional biological definitions of the individual. We were very taken with the biological work we read on symbiotic, multispecies biology. After all, we are technically more bacteria than *Homo sapiens*! We have multiple species in our bodies and yet we write them off in typical gendered and classed term as 'housekeepers' and support staff not worth considering. The master-mind 'brain' gets all the credit, as usual! One line of our work has taken seriously the attempt to decenter the human and think more holistically about biological evolution. It has been exciting!"

"Can you imagine—developing a story where non-Darwinian evolution, non-Mendelian inheritance, challenging the 'sex' and the 'individual' as categories all at once? Created quite the storm. But, you know, once you have a broad training and an open mind, it is so much easier to identify the myopia and blind spots of disciplines. Suddenly you observe things that are really staring us in the face—if we are open to recognizing them. It is cool to be the stars in a sea of white male scientists," said Tabrez gleefully. "But you know, if not for our friendship, if not for our diverse backgrounds, if not for the community, and if not for us talking about and teaching each other wide range of fields and thinkers—Foucault and Darwin, literature, gender and queer studies, postcolonial studies, geology, biology, math, biochemistry, statistics, and meteorology—we would never have come upon this amazing discovery."

"The summer with the researchers really planted so many seeds. We saw the futility of disciplinary silos. We've never forgotten that," said Tara.

"Now, that is a requirement in all our work," laughed Tabrez.

"We also work on integrative medicine. You see, we realized early on that it would not be easy for us to continue to work in this little town. And we really love it here. We also would all like to continue to work together. We find the western model of academic tenure and mentorship such an intellectually inbred model that closes rather than opens the mind. Now, through this institute we are entirely self-funded. And all this we owe to Tara's puttering around," said Tulasi.

"Yeah, I was really taken with the Mantri of Tantri. And so I began to read more about alternative medicine in the west. It was funny because we have such a rich tradition. And while Indians ignored it, pharmaceutical companies were increasingly exploring local traditions and medicines, then finding the indigenous plants, patenting the active ingredient. Ultimately, locals could not afford that medicine! I thought, why can't we do it ourselves? Bring scrutiny to alternate medicine."

"So we began a center for integrative medicine here. Over there," she pointed to various buildings, "are the labs. The facilities are owned by the town. It's a complex co-operative in structure. Local knowledge through oral history, traditions in various communities, local medical systems have proved invaluable in discovering new treatments and systematizing this knowledge. Very often we discover that the answer isn't in just a little 'pill' but a way of life. It is this holistic conception of the body and health we are trying to promote. But as you can see, we continue to work with morning glories. We've retained those origins. Tara also found some medicinal uses for morning glory roots. They are bronchodilators and help with asthma and other respiratory illnesses."

"And there's always the seeds!" said Tulasi with a mischievous look about her.

"And there you have it," said Tabrez. "You know it's been so much fun. I don't understand how people work alone, stay stuck in their labs, write individually. I'd die of boredom. The three of us have had such a blast, it's hard to communicate that adequately."

And so we leave the Thirumbaram Three as they continue laughing, Tara's hair blowing with the wind, Tulasi twining a plant around its stake, and Tabrez clowning around. They walk by the now famous field. The morning glory buds are full, ready to bloom early next morning. They climb up their favorite tree and sit down on each of their perches and watch the sun go down.

PART II

Geographies of Variation
The Case of Invasion Biology

Fitcher was saddened to learn that his own compatriots were responsible for the garden's decline . . . Fitcher's revulsion at the sight of the wilderness that loomed before him now: the untrimmed crowns of the garden's trees had grown into each other, forming a canopy so dense that the grounds beneath, with their flower-beds and flag-stoned pathways, were shrouded in darkness; along the peripheries of the compound, the greenery was as impenetrable as a wall, and the unclipped aerial roots of the banyans that flanked the main gateway had thickened into a forbidding barrier—a portcullis that seemed to be designed to keep intruders at bay. This was no primeval jungle, for no ordinary wilderness would contain such a proliferation of species, from different continents. In Nature there existed no forest where African creepers were at war with Chinese trees, nor one where Indian shrubs and Brazilian vines were locked in mortal embrace. This was a work of Man, a botanical Babel.

—Amitav Ghosh, *River of Smoke*

CHAPTER FOUR

Alien Nation

A Recent Biography

Since we have all been more or less constantly on the move
since our ancestors decamped from the old neighborhood in
Ethiopia, 195,000 years ago, you'd think that, as a species,
we might have worked through our hostility and suspicion of
newbies by now. But we haven't.

—Peter Behrens, *The New York Times*, March 16, 2012

Give me your tired, your poor, Your huddled masses yearning
to breathe free, The wretched refuse of your teeming shore.
Send these, the homeless, tempest-tossed, to me: I lift my
lamp beside the golden door.

—Emma Lazarus, *The New Colossus*

A nation that cannot control its borders is not a nation.

—Ronald Reagan on signing the Immigration Reform and
    Control Act, 1986.

Where do plants, animals, and humans originate, and does it matter? Who
migrates where, when, and why and to what consequences? Part II explores
the naturecultural worlds of invasion biology. This project on biological invasions began simply enough as an interdisciplinary project in science and the
feminist studies of science. This collaborative project was provoked by the
growing public alarm about exotic and foreign species. As a biologist, I was
keen to understand and explain the shifting landscape—why are numbers of
native species dwindling and foreign species increasing? From feminist studies,
I was interested in our generic fear about the "foreign" and invasion biology's

articulations of space and belonging, and of how science comes to mobilize a public toward social and environmental policy. More centrally, how did biology come to its categories of analysis, how did the production of knowledge in science proceed? The project continued as both the biologist and cultural studies scholar in me watched, observed, and analyzed the evolving collaboration.

In this chapter I trace the co-production of knowledge on the global migrations of biota—humans, plants, and animals—by exploring a brief *bio*graphy of life in the United States over the past decade. The following chapter explores my collaborative field experiments on invasive species, and the final chapter in this part is a meditation on my growing identification of being an alien studying aliens. How might such subjectivities, this spirit of kinship and alliance, be a source of productive knowledge production?

At the outset, I should be clear about my political and ecological sympathies. To me, humans are a part of nature and have co-evolved with the flora and fauna. A return to a pre-human world is not possible or even desirable. I do not believe that nature was once good, pure, and virginal and would have remained so without humans. Nature without humans was and is not a static idyllic utopia. Instead, to me, nature is a concept, an idea, and a place that is co-produced through the interactions and entanglements of various organisms, histories, and geographies. To me many of the issues we debate are largely moot: Is globalization good or bad? Should our borders be open or closed? Is hybridity or multiraciality a good or bad thing (Cardozo and Subramaniam 2013)? History is very illustrative—globalization has long been here, our borders have always been porous, and multiraciality and multiple sexualities are not new. Histories of our Planet remind us that complex ecological relationships produce and have always produced a dynamic, ever-changing, and complex ecological system. Nature is ever evolving—we see periods of little change, or stasis, and those with tremendous flux.

While I may not subscribe to a nostalgic pure nature of yesteryears, it is fitting to call our contemporary epoch the *Anthropocene*, an era driven primarily by the impact of human actions (Crutzen and Stoerme 2000). From colonial powers that reshaped local ecosystems to maximize resource extraction to neocolonial modes of contemporary resource attraction, from human-made war and poverty that have decimated many populations (human and other) to world social and economic systems that have enslaved and impoverished entire nations and their biota, human influence is profound. Refusing the unproductive choices of a nostalgic past or an anarchic future, I turn to a naturecultural vision of responsible and ethical living with our cohabitants, a vision that is always politically astute and reflexive of the complex histories of gender, race, class, sexuality, and nation that have shaped our ideas of nature and the natural.

Variation, Diversity, and Adaptation

Geography, it turns out, is central to the idea of variation in biology as we understand it today. Darwin credited his many geographical explorations, especially his voyage on the HMS *Beagle*, as critical to his developing theories, as was the tremendous variation he observed within and across geographies. Geography also has another important meaning that is central to our theories of evolution, that is, place. Variation is the raw material for natural selection and for geographies and their environmental contexts that lead to adaptations of organisms and the actions of natural selection. In Darwin's vision, nature is neither static nor stable. Organisms evolve as evolutionary contexts change. In addition to geological ruptures that literally move bodies of land, nature facilitates movements in many ways. Organisms migrate—a seed can travel tucked into a bird's feather or an animal's fur; an organism, egg, or seed can be carried on a floating log; strong winds carry pollen and seeds afar. Organisms housed within other organisms (symbionts, parasites, and other co-evolved entities) move afar with their hosts. Yet, ideas of nature *and* place go hand in hand—humans, plants, and animals come to "belong" to particular geographies, and we have come to talk about "nature in place" and "nature out of place."

Nature's Place

The idea of invasion is predicated on a discourse of "nature in place" and "nature out of place"—and by definition invasive species are introduced species that do not belong. This idea of "nature in place" has a complex and nonlinear history. The concept of "nativeness" was first introduced by the English botanist John Henslow in 1835, and subsequently Hewett Watson used this to delineate "a true British flora" (Davis et al. 2011a, Chew and Hamilton 2011). Watson's terminology drew on English common law about human citizenship rights (Chew and Hamilton 2011). Of course, "true" British flora constitutes the "not true" and the now familiar binary of the native/alien emerged. While the term continued to be used in the coming decades, no general policy about native/aliens emerged (Davis et al. 2011a). In our recent genealogy of invasion biology, Charles Elton's 1958 *The Ecology of Invasions* is cited as the classic book that ushered in the field of invasion biology. While laws policed biotic flows, it was only with SCOPE (Scientific Committee on Problems of the Environment) in 1983 that "invasion biology" as a field or discipline emerged (Davis et al. 2011b). Today, according to the U.S. Department of Agriculture (USDA), "invasive plants are introduced species that can thrive in areas beyond their natural range of dispersal. These plants are characteristically adaptable, aggressive, and have a high reproductive capacity. Their vigor combined with a lack of natural enemies often leads to outbreak populations" (U.S. Department of Agriculture 2012). Indeed, in recent decades, work on invasive plant species has exploded. The frenzied alarm has been sounded by groups of the right and left, environmentalists and non-environmentalists alike. At the level of research and policy, this is a fertile area. The USDA, state governments, National Science Foundation committees, as well as environmental groups such as the Nature Conservancy and the Sierra Club all have invasive species programs.

We have historically imagined our relationship with the biota of the world in numerous and diverse ways. In his influential book *Ecological Imperialism*, Alfred Cosby argues that the roots of Europeans' domination of the western world lie in their creating "New-Europes" wherever they went, especially in North and South America, Australia, and New Zealand (Crosby 1986). Rather than thinking of European domination as the result of technology, Crosby argues that we should understand it as simultaneously biological and ecological. Where Europeans went, their agriculture and animals went; they thrived while indigenous ecosystems collapsed. This vast migration of species ushered in a bioinvasion of mass proportions by the conquerors' animals, plants, weeds, and germs, yielding a "great reshuffling" (Crosby 1986, McNeely 2001, Warren 2007, Weiner 1996). Some plants were now ubiquitous; as Crosby remarks, "the sun never sets on the empire of the dandelion" (Crosby 1986: 7). The science of breeding and horticulture led to scientific breeding stations around the globe that turned raw materials from the colonies into plantation crops for the British, French, and other colonies and empires, including the Americas (Brockway 2002).

As early as 1776, the United States dealt with its first insect invasion by the Hessian fly. While the Americans and British responded with nationalist and xenophobic fervor, these invasions were seen as isolated events (Pauly 2002). Indeed, in the late nineteenth and early twentieth centuries, the USDA had an active program where biologists as "explorers" roamed the globe in search of new and interesting plants of economic and aesthetic interest. For example, Dr. Douglas Fairchild, director of USDA's Section of Seed and Plant Introduction from 1898 to 1928, is said to have introduced more than 80,000 species and varieties into the United States (Conlon 2010). Likewise, the American Acclimatization Society introduced a variety of plants and animals, including an attempt to introduce all of the bird species mentioned in Shakespeare's works to New York City's Central Park in the 1890s (Blair 2008). Such an open and laissez faire U.S. policy ended late in the nineteenth century, however, due in part to America's changing relationship with nature. In the decades after the Civil

War, industrialization, urbanization, and westward expansion transformed the nation's landscapes and redefined Americans' relationship with nature (Rone 2008). This new love of nature was evidenced in the dramatic growth in the number of Americans who considered themselves "nature lovers," and Americans saw their love of nature as *the* quality that distinguished the "natives" from the new immigrants. A love for nature translated into a zeal to protect nature, and immigrants came to be seen as not loving nature and as the problem. Nativists increasingly challenged federal government passivity about immigration. The "native" emerges as site of "purity" in our conceptions of humans and plant and animal ecologies. Indeed, as Philip Pauly (1996) notes, the paradigm of the nativist approach was the Chinese Exclusion Act, passed at the insistence of California workingmen in 1882, a year after the state's quarantine law. After World War I, Congress introduced limitations on entries of all European immigrant groups through the Immigration Act of 1924.

During the hearings for the Immigration Act, eugenics played a big role. Experts testified on the importance of bringing "good stock" into the United States, warning of the public cost of "bad stock" in caring for those who had physical and mental disabilities or diseases (Allen 1996). The new demographic policy on limiting human immigration was a central part of the vision of scientific bureaucrats to make the United States ecologically independent (Pauly 1996). Eugenic concerns span variation's genealogies and geographies—and fears of the high reproduction of the "poor" stock directly fueled concerns about the erosion of the environment (Stern 2005). While evolutionary biology and eugenics are intimately linked, so are environmentalism, ecology, and eugenics.

Like human immigrants, the arrival of foreign plants and animals is deeply intertwined with the needs of the U.S. economy—as ornamentals, recreation, utility, food, comfort, entertainment, soil erosion, and pest or weed control. The Asian carp was imported as a "worker fish" to clean up areas with aquatic weeds (Burdick 2005, Frazier 2010), and kudzu was promoted by the government to prevent soil erosion (Burdick 2005, Frazier 2010). More than half of the plants known to be invasive in North America were originally imported for their horticultural use (Marinelli and Randall 1996). Human "dispersal" is key for the movement of plants and animals (Bean 2007, Chew and Hamilton 2011, Klein 2002, Pysek et al. 2004). The majority arrived at our behest because *we* invited them; these invasions are best understood as *invited invasions* (Cardozo and Subramaniam 2013). These invasions were proliferated by the domestication, acclimatization, and breeding sciences—scientific disciplines that converted "foreign" species into economically valuable species (Brockway 2002). In the same spirit, botanical gardens introduced U.S. citizens, especially children, to many species not native to the United States, ushering in a long history of fascination with exotic plants from across the globe (Daston and Park 2001).

The impact of humans on the rest of the environment has been a critical focus in recent environmentalist thought and activism. The profound migrations of people, plants, and animals have occurred within the complex circuits of colonialism, tourism, forced and voluntary migrations, trade, and globalization. A recent Gallup World Poll suggests that 1.1 billion people, or one-quarter of Earth's adults, want to move temporarily to another country to find more profitable work. Another 630 million people would like to move permanently (Matt 2012). While humans migrate, landscapes of home are always reinvoked in new geographies, as migrants have historically turned to the comfort of landscapes that remind them of home. Indeed, with the advent of cars, ships, and planes, and trade and travel, there is virtually no region of the Planet unconnected with another. World biogeography is constantly made and remade through the varied circuits of history and politics. Seemingly innocent ideas such as biological variation become entangled and translated into complex ideas of difference. Which landscapes are superior and which are inferior? What constitutes home and abroad? Who is deemed native and alien?

Naturecultural Migrations

Understanding the geographies of variation through a naturecultural lens is productive. Environmentalists raise the alarm over the growing number of invasive species and anti-immigration activists raise the alarm about immigrants. Unbeknownst to each, the anti-invasive and the anti-immigrant rhetorics help feed a rampant xenophobia about the problem with these "outsiders." Welcome to the naturecultural world! We cannot understand plant, animal, or human evolution without understanding their connections to each other. Yet biologists study plant immigration and social scientists human immigration, usually on different geographical ends of a campus. Environmental and immigrant activists likewise exercise their influence often through entirely different networks of influence and circuits of power. Yet, I would contend, that while the natural and political worlds appear separate and are reported in different sections of the newspaper and studied in different disciplines, xenophobic sentiments strengthen as they travel across campaigns against foreign plants, animals, and humans.

The power of the framework of naturecultures is in illuminating the complex circulation of knowledge. Indeed, the categories of native and exotic, foreign and alien, which seem to be such "human" categories, have their origins in distinctions in early plant cultivations to explain the distribution of species—in distinguishing between the "wild" and the domesticated "garden" (Grese 2011). It is the failure to see naturecultures that leads to one of the great ironies where in an era of globalization, there is a renewed call for the importance of the "local" and the protection of the indigenous. The alarm about human immigration is accompanied by an alarm about alien plants and animals. With the increased permeability of nations and their borders,[1] and the increased consumption and celebration of our common natures and cultures, we begin to obsess about our different natures and cultures with a fervent nationalism, stressing the need to close our borders to "outsiders." The anxieties around the free movement of capital, commodities, entertainment, and the copious consumption of natural and cultural products have reached fever pitch. In the realm of culture and the economy, nationalisms, fundamentalisms, World Trade Organization (WTO) protests, censorship of "foreign" influences, and calls for the preservation of national cultures abound.[2] The globalization of markets, and the real and perceived lack of local control, feed nationalist discourse. High unemployment rates, coupled with the search of companies for cheap labor abroad and the easing of immigration into the country, have increasingly been perceived as threats to local employment. These shifts continue to be viewed by some elements of both the right and the left as a problem of immigration.

The histories and interconnected political valences (in biological and social histories) of humans and plants and animals are entirely obscured and lost. Yet through a naturecultural lens, they come crisply into focus. Suddenly, we begin to see that our anxiety about a fast-changing Planet, the immense mobility and flux of our lives, and the fear of anarchy and chaos appear in a myriad of sites. Rhetoric of "nature out of place" is everywhere—among economists, labor activists, unions, politicians, border control agents, environmentalists, biologists, and epidemiologists. We feel under siege, an illusion of multiple data points of multiple phenomena, all pointing to the same problem—nature "out of place." Suddenly these anxieties seem right, obvious, commonsensical, and, above all, "natural." We respond with xenophobic ideologies across multiple sites, trying to restore nature to its rightful place. What a naturecultural analysis allows us to see is the circulation of knowledge—that these are not different data points after all, but the same data points appearing in multiple loci. Indeed, these are all symptoms of the same anxieties, the same problem—our cultural anxieties of a purportedly fast-changing world, anxieties that we and "our kind" will be left behind.

Science, like politics, is not monolithic. While certain stories and viewpoints dominate, there are *always* other viewpoints. Science is only one site in our naturecultural world where social anxieties manifest themselves. As politicians, activists, and a society at large grapple with ideas of the "nation" and who belongs, these same battles spill over into the scientific and natural worlds. The case of invasion biology is an ideal case in point. If the ideological battles are being waged across multiple sites, it also suggests that a naturecultural analysis allows a fertile opportunity for resistance and collaboration.

Alien Nation! A Recent *Biography*: A Tale in Four Acts

Thinking natureculturally reminds us that plants, animals, and humans share a common *bio*graphy. In examining the history of foreign plants, Philip Pauly (1996: 70) argues that

> commonplace symbolic connections between geographically identified organisms and humans were omnipresent and powerful. Americans perceived the English sparrow as an avian Cockney pushing aside larger but better-mannered American birds. The gypsy moth's devastation of respectable neighborhoods confirmed casual prejudices about Gypsies. The introduction of pulpy tropical fruits, carrying the aromas of the Far East, really did increase sensuality among Americans raised on tart apples and bland pears. These specific associations could operate at high cultural levels, and their implications could be molded consciously.

I explore the ghostly naturecultural world of the "foreign" by examining invasion biology in four "Acts," four moments in the popular media over the past decade, each demonstrating the multiple cultural valences on immigration. The stories come from a wide array of players, from scientists and nonscientists, environmentalists and nonenvironmentalists alike. While I focus on popular news sources here, the spirit of the debate is well evident in the scientific literature as well. *Act I: The Aliens Have Landed* explores the explosion of research on invasion biology in the 1990s, the decade when invasion biology emerged as a discipline of its own (Davis et al. 2011a). During that time, the efforts of biologists and environmental activists dealing with alien biota were reported every day in newspapers and magazines across the country. This section also sets up the basic arguments and rhetoric that frame invasion biology. Over the following decade, the daily reporting on invasive species continued. The following three acts represent departures from these daily reporting—moments that gave me pause or cause to sit up and wonder, "This is different! What is going on?" *Act II: And Then the Towers Fell* examines one of several cases of sensationalized mass hysteria around particular aliens, in this case the Chinese snakehead in Crofton Pond in 2002. The case enacts the arc of the politics of a post–September 11, 2001, world. *Act III: An Obligation of Reluctance*, set in 2006, features an op-ed in *The New York Times* unreservedly and exuberantly promoting an embrace of the wild diversity of all plants, including alien biota. The piece accuses those wanting to kill weeds and other biota of rampant xenophobia. Finally, *Act IV: The Invasive Species War* examines a well-publicized debate in 2011 in *Nature* and subsequently reported in the popular press of a debate among biologists on the question of invasive species.

A Note on Studying Naturecultures

Why focus on popular media and popular representations? This work is about tracking the production of knowledge. Take two parts in the process of the production of knowledge. First, biologists produce knowledge about invasive species and their destructive impact and publish these in scientific journals. Second, environmental societies across the country mobilize private citizens to sacrifice their weekend leisure in joining campaigns to uproot invasive plants from their local habitats (Neyhfakh 2011). How do we connect the two? To understand this, we need to understand the knowledge biologists produce; who biologists are; how biologists garner funding for their work; how work gets published; and how biologists promote their work to the nation, lobby popular support, and shape environmental and public policy. First, it is not as though biologists do their work and nonbiologists "apply" the work. The two are inextricably interconnected. Popular media is grounded in the work of invasion biology, often involving the same biologists. But only selected research gets the attention of the popular press. Scientific jargon and complexity is often translated and made accessible—and some things are often lost in the translation. So while we need to be nuanced about how knowledge and theories travel, they are indeed connected and do indeed travel. Second, contrary to popular mythology, most scientists are not bearded rationalists plotting the latest bomb! Rather, even a brief perusal of biology or any of the sciences demonstrates a history of intellectually curious, politically engaged, passionate, and concerned scientist-citizens. Scientists are located in society, are embedded in cultures, and bring to their work a diverse range of cultural understandings, varied politics, and ambivalence present in society.[3] Indeed, scientists—as is evident in most invasion biologists presented here—are deeply involved in influencing environmental policy, in getting the public and politicians to take what they see as the problem of environmental management seriously. This is a field where the scientists have taken very vocal and public stances, and not a case where their ideas are being simplified or misrepresented. Finally, science would not function without funding. Institutions of science are perennially working to garner more funds and showcase the importance and impact of their work. They do this through lobbying the government and private sector for funds, but, most important, in making visible the public value of their work. The public face of science is critical to the institution of science. Fear narratives have been a potent force in mobilizing politicians and the public (B. Hartmann et al. 2005). Through all these factors, nature and culture co-constitute scientific knowledge.

But there is a deeper philosophical problem here. One of the challenges in talking about naturecultures is that it requires a different set of tools than we have. Our theoretical frameworks, our vocabulary, categories, rhetoric, and discourses, are all geared to an academy with individual disciplinary histories and traditions. Do humans belong to the natural or cultural worlds? As an animal species with a common evolutionary history that links all of life, humans are clearly part of the natural world. And indeed, biologists and biological anthropologists study humans and their impact as part of "nature." But as creators of a complex society that has intruded on most aspects of the Planet, we humans are clearly part of the cultural world as well. And indeed, the disciplines of the social science and humanities are devoted to such explorations. Thus, studying humans poses epistemological and ontological problems and, of course, many disciplinary confusions. Further confusing the issue, plants and animals are considered by some to be part of the cultural world as well.

We also have to confront the perennial frame of the long-enduring debate: matter versus representation. When we talk about how humans *are* and what humans *do*—being and knowing—we have to employ language, be they words (visual/oral), pictures, graphs, or numbers. Each has its own particular and peculiar histories. When we talk about "alien" species, is the problem just about "representation" or a problem with terminology? Let us, for example, say that we create a new term *XXXX* to talk about foreign plants. Do we solve the problem? No! The fundamental point of thinking natureculturally is understanding that so long as the category of "foreign species" (whatever term we may call it) exists in our minds, it is still linked biologically, rhetorically, historically, and philosophically to a binary world of natives/aliens. Some have suggested that we use different and less pejorative words, such as *piggy backers, opportunists, spawn, mirrors, providers, hybrids, tricksters, matrices, transients, founts,* or *teachers* (Larson 2007b). The point is not only the xenophobia that permeates the terms of invasion biology, as some have argued, but also the way natives and aliens are presented as a biologically and ecologically useful binary. This is not just about language or words but the ways in which human history has created particular categories or "ontologies." The point of thinking natureculturally is to understand that whether we deploy visual, tactile, auditory, or linguistic vocabularies, these are all languages with shared histories, categories, and structures. In fact, the neurophysiologist Richard Gregory estimates that visual perception is more than 90 percent memory and less than 10 percent nerve signals (1998). We can never escape these cultural, historical, and disciplinary memories or the ghostly hauntings of naturecultures. It misrepresents the problem and misidentifies the solution. Understanding them as naturecultures allows us to step back to see what is underpinning these natural and cultural battles—a profound anxiety and deep ambivalence about our ideas of the nation, home, and belonging.

ACT I: THE ALIENS HAVE LANDED

When I began thinking about this project in the late 1990s, an article in a recent special issue on "Biological Invaders" in the prestigious journal *Science* began as follows:

> One spring morning in 1995, ecologist Jayne Belnap walked into a dry grassland in Canyonlands National Park, Utah, an area that she has been studying for more than 15 years. "I literally stopped and went, 'Oh my God!'" she recalls. The natural grassland—with needle grass, Indian rice grass, saltbush, and the occasional pinyon-juniper tree—that Belnap had seen the year before no longer existed; it had become overgrown with 2-foot-high Eurasian cheatgrass. "I was stunned," says Belnap. "It was like the aliens had landed." (Enserink 1999: 1834)

While I was reading the literature to understand the ecologies of native and exotic species, the biological and cultural work showed a growing panic about alien and exotic plants and animals. Newspaper articles, magazines, journals, and websites all demanded urgent action to stem the rise of exotic biota.

"They Came, They Bred, They Conquered"

Newspapers and magazines introduced the topic of biological invasions with the sound of alarm. Consider some of the headlines:

New Rules Seek to Prevent Invasive Stowaways (Barringer 2012)
Alien Invasion: They're Green, They're Mean, and They May Be Taking
    Over a Park or Preserve Near You (Cheater 1992)
The Invasion of the Woodland Soil Snatchers (Barbara Stewart 2001)

Native Species Invaded (United Press International 1998)
Bio-invasions Spark Concerns (*CQ Researcher* 2000)
It's a Cancer (Weaver qtd. in Verrengia 1999)
Creepy Strangler Climbs Oregon's Least-wanted List (Brinckman 2001)
Biological Invaders Threaten U.S. Ecology (K. A. McDonald 1999)
U.S. Can't Handle Today's Tide of Immigrants (Yeh 1995)
Alien Threat (Bright 1998)
Biological Invaders Sweep In (Enserink 1999)
Stemming the Tide of Invading Species (Kaiser 1999)
Congress Threatens Wild Immigrants (Weiner 1996)
Invasive Species: Pathogens of Globalization (Bright 1999)

What is striking is that the majority of these headlines do not specify that the article is about plants and animals but rather present a more generalized classic fear of the outsider, the alien that is here to take over the country. An opening line of an article reads: "The survey is not even halfway done, yet it has already revealed a disturbing trend: immigrants are forcing old-timers out of their homes" (B. Stewart 2001: 1B). Invaders are reported to be "racing out of control," causing "an explosion in slow motion" (Hebert 1998). Aliens, they claim, are redrawing the global landscape in ways no one imagined. Exotic plants, they argue, are irreversibly altering waterways and farmlands. The "irreversibility" is highlighted as a way to stress the sharp departure from the past—a vision of how we are moving from a peaceful, co-evolved nature in perfect harmony and balance to an uncertain future with alien and exotic plants and animals. They argue that we cannot recapture the glorious past, our nostalgia for a pure and uncontaminated nature in harmony and balance, if we do not act *now* to stem the tide of outsiders.

A careful examination of the biological and cultural literature reveals that the parallels in the rhetoric surrounding foreign plants and those of foreign peoples are striking. As Nancy Tomes has argued, our anxieties about social incorporation (associated with expanding markets, increasingly permeable borders and boundaries, growing affordability of travel, and mass immigration) have historically spilled into our conceptions of nature. For example, she documents how our panic about germs has historically coincided with periods of heavy immigration to the United States, of groups perceived as "alien" and difficult to assimilate. She documents these germ panics in the early twentieth century in response to the new immigration from Eastern and Southern Europe and in the late twentieth century to the new immigration from Asia, Africa, and Latin America. "Fear of racial impurities and suspicions of immigrant hygiene practices are common elements of both periods," she writes. "These fears heightened the germ panic by the greater ease and frequency with which immigrants travel back and forth between their old, presumably disease ridden countries and their new, germ obsessed American homeland" (Tomes 2000: 195). Like these earlier germ panics surrounding immigration and immigrants, questions of hygiene and disease haunt exotic plants and animals. Similar to the unhygienic immigrants, alien plants are accused of "crowd[ing] out native plants and animals, spread[ing] disease, damag[ing] crops, and threaten[ing] drinking water supplies" (Verrengia 2000). The xenophobic rhetoric that surrounds immigrants is extended to plants and animals.

The first parallel is that aliens are "other." One *Wall Street Journal* article quotes a biologist's first encounter with an Asian eel. "The minute I saw it, I knew it wasn't from here," he said (Robichaux 2000: 12A). Second is the idea that aliens/exotic plants are everywhere, taking over everything: "They're in national parks and monuments. In wildlife refuges and coastal marine sanctuaries. In wilderness areas that were intended to remain living dioramas of our American paradise lost" (Verrengia 2000). "Today, invasive aliens afflict almost every habitat in the country, from farms and pastures to forests and wetlands—and as every homeowner knows, gardens, flower beds and lawns" (Cheater 1992: 24).

The third parallel is the suggestion that they are silently growing in strength and numbers. So even if you haven't noticed it, be warned about the alien invasion. If you haven't heard about biological invasions, it is because "invasion of alien plants into natural areas has been stealthy and silent, and thus largely ignored." E. O. Wilson states: "Alien species are the stealth destroyers of the American environment" (qtd. in K. A. McDonald 1999: A15). Articles remind us that alien plants are "evil beauties"—while they may appear to look harmless and even beautiful, they are evil because they destroy native plants and habitats (Cheater 1992). These campaigns are eerily similar to the antiprostitution public health propaganda put out to soldiers. The fourth parallel is that aliens are difficult to destroy and will persist because they can withstand extreme situations. In an article on the invasion of the Asian eel in Florida, the author notes the following:

> The eel's most alarming trait, though, is its uncanny ability to survive extreme conditions. In one study by a Harvard zoologist, an Asian swamp eel lived seven months in a damp towel without food or water. The olive-brown creature prefers tropical waters, yet it can flourish in subzero temperatures. It prefers fresh water but can tolerate high salinity. It breathes under water like a fish, but can slither across dry land, sometimes in packs of 50 or more,

sucking air through a two-holed snout . . . Even more of a riddle is how to kill the eel: It thus far appears almost immune to poisons and dynamite. (Robichaux 2000: 12A)

The fifth parallel is that aliens are "aggressive predators and pests and are prolific in nature, reproducing rapidly" (Verrangia 2000). They are the "new vermin" of modernity (Baskin 2002, Smout 2003). This rhetoric of uncontrollable fertility and reproduction is another hallmark of human immigrants. Repeatedly alien plants are characterized as aggressive, uncontrollable, prolific, invasive, and expanding. One article summarized it as "They Came, They Bred, They Conquered" (Bright 1999). Alien species are characterized as destroyers of everything around them. A park warden is quoted as saying, "To me, the nutria [swamp rats] are no different than somebody taking a bulldozer to the marsh" (Verrengia 2000).

Sixth, once these plants gain a foothold, they never look back (Cheater 1992). Singularly motivated to take over native land, aliens have become disconnected from their homelands and will never return and are, therefore, "here to stay." Finally, like human immigrants, the greatest focus is on their economic costs because it is believed that they consume resources and return nothing. "Exotic species are a parasite on the US economy, sapping an estimated $138 billion annually, nearly twice the annual state budget of NY, or a third more than Bill Gates' personal fortune" (Verrengia 1999).

Not only are aliens invading rural and natural habitats, they are also endangering the cities. "Cities invaded," articles cry. From historical sites to urban hardwoods, alien insects are reported to be causing millions of dollars worth of damage (Verrengia 1999).

> Just as human immigrants may find more opportunities in an already over-crowded city than in a small town, invasive plants take advantage of the constant turnover and jockeying for position that characterizes species-rich ecological communities. The classical dictum that "diversity begets stability," Stohlgren says, is simply not true in some ecosystems. Communities with high diversity tend to be in constant flux, creating openings for invasives. From a conservation perspective, the results of these multi-site, multi-scale studies are disturbing. The invasions may threaten some of the last strong-holds of certain biologically rich habitats, such as tall-grass prairie, aspen woodlands, and moist riparian zones. (U.S. Geological Survey 1999)

Finally, we see foreign plants, animals, and humans as exploited and exploitable. In drawing attention to the political and ethnic category Asian American, Karen Cardozo and I argue that we ought to understand the term *Asian/ America* as a multispecies *assembling*. Humans, plants, and animals share incredibly intertwined histories. For example, foreign humans come into this land because of the "need" for labor, expertise, skills, or beauty at one point in history only to be severely regulated once they are here, and then rejected, shunned, and sent back once they have outlived their usefulness. We see this in Chinese railroad workers, migrant workers in the farms, mail order brides, technology workers in the 1990s, and of course doctors and nurses (who are still in short supply). Similarly, the now invasive Asian carp was brought into the country as a "worker fish" to help clean algae-filled canals, and kudzu was once promoted by the U.S. government, who paid farmers to plant it for erosion and soil control. Yet these histories get entirely forgotten in the xenophobic quest to eradicate individual ethnic groups or individual species to bolster the idea of a pure nature and nation eliminating its foreigners (Cardozo and Subramaniam 2013).

The Oversexed Female

One of the classic metaphors surrounding immigrants is the oversexualized female. Foreign women are always associated with superfertility—reproduction gone amok. Such a view suggests that the consumption of economic resources by invaders today will only multiply in future generations through rampant overbreeding and overpopulation. Consider this:

> Canada thistle is a classic invasive. One flowering stem can produce as many as 40,000 seeds, which can lie in the ground for as long as 20 years and still germinate. And once the plant starts to grow, it doesn't stop. Through an extensive system of horizontal roots, a thistle plant can expand as much as 20 feet in one season. Plowing up the weed is no help; indeed, it exacerbates the problem; even root fragments less than an inch long can produce new stems . . . The challenge posed by thistle is heightened because, like other troublesome aliens, it has few enemies. (Cheater 1992: 27)

Along with the super-fertility of exotic/alien plants is the fear of miscegenation. There is much concern about the ability of exotic plants to cross-fertilize and cross-contaminate native plants to produce hybrids. Campaigns against the foreign are often constructed around preserving the "genetic integrity" of native species and the prevention of their interbreeding with native populations (Smout 2003). Native females are, of course, in this story, passive, helpless victims of the sexual proclivity of foreign/exotic males.

Responding to Alien Species

Journalists and scientists borrow the same images of illegal immigrants arriving in the country by means of difficult, sometimes stealthy journeys, when they describe the entry of exotic plants and animals. Alien plant and animal movements are described with the same metaphors of illegal, unwelcome, and unlawful entry (Hebert 1998). The response to such unlawful and stealthy entrants parallels our immigration policy with changing rules and images of armed guards patrolling borders (Barringer 2012). Former Montana Governor Marc Racicot is passionate: "I just hate them. They are genetically deviant miscreants that have no rightful place on this Planet. We all have to be a part of this war on weeds" (Associated Press 1999 qtd. in Hettinger 2011: 194). Interior Secretary Bruce Babbitt of the Clinton administration called the "alien species invasion" an "explosion in slow motion" that was turning even "staunch conservationists into stone killers" (Verrengia 1999). Like immigration and the drug problems, the language called for a need to "fight" and wage wars against exotic/alien plants and animals (Larson 2005). In 1999, President Clinton signed an executive order creating the national Invasive Species Management Plan directing federal agencies to "mobilize the federal government to defend against these aggressive predators and pests" (Hebert 1998). Thus the "Feds" were called on to "fight the invaders" and defend the nation against the "growing threat from non-native species" (Hebert 1998). In rather strong language, then Interior Secretary Bruce Babbitt summarized the situation: "Invasive alien species . . . homogenize the diversity of creation . . . Weeds—slowly, silently, almost invisibly, but steadily—spread all around us until, literally, encircled, we can no longer turn our backs. The invasion is now our problem, our battle, our enemy . . . we must act now and act as one [in order to] beat this silent enemy" (Babbitt 1998).

A review sponsored by the Ecological Society of America published in the *Issues of Ecology* concluded that the current strategy of denying entry only to species already proven noxious or detrimental should change (Mack 2000). Instead of an "innocent until proven guilty" stance, we should instead adopt a "guilty until proven innocent" one. This strategy is further racialized when a biologist rephrases this by suggesting we replace a "blacklist" (where a species must be proved to be harmful before banned) with a "whitelist" (where species has to proved to be safe before entry) (Simberloff qtd. in Todd 2001: 253). Paralleling the language of third world debt that marks colonial legacies in the human world, we have the "invasion debt" that recognizes the legacies of biotic movements and accounts for the long list of foreign species that are already in the country (D. M. Richardson 2011). In all of this, exotic and alien plants are marked as "guilty," foreigners, and black and, therefore, kept out purely by some notions of the virtue of their identity.

Natives

What is tragic in all this is of course the impact on the poor natives. "Native Species Invaded," "Paradise Lost," "Keeping Paradise Safe for the Natives" are the repeated cries. Native species are presented as hapless victims that are outcompeted and outmaneuvered by exotic plants. Very often, exotic plants are credited with (and by implication, native species are denied) basic physiological functions such as reproduction and the capacity to adapt. For example, "When an exotic species establishes a beachhead, it can proliferate over time and spread to new areas. It can also adapt—it tends to get better and better at exploiting an area's resources, and at suppressing native species" (Bright 1999).

Invaders, Interior Secretary Bruce Babbitt claimed, "are racing out of control as the nonnative species in many cases overpower native species and alter regional ecosystems" (qtd. in Hebert 1998). Not only do they crowd out native plants and animals, but they also endanger food production through the spread of disease and damage to crops and they affect humans through threatening drinking water supplies. Consider this:

> English ivy joins 99 plants on a state list of botanical miscreants that includes Himalayan blackberry, Scotch thistle and poison hemlock. With dark green leaves and an aristocratic heritage, however, it looks like anything but a menace.
> Don't be fooled.
> The creeper loves Oregon, where it has no natural enemies.
> It needs little sunlight. It loves mild, wet climates.
> Robust and inspired, English ivy jumps garden borders, spreading across forest floors, smothering and killing ferns, shrubs and other plants that support elaborate ecosystems and provide feeding opportunities for wildlife. Insatiable, English ivy then climbs and wraps trees, choking off light and air. (Brinckman 2001: 1A)

Articles invariably end with a nostalgic lament to the destruction of native forests and the loss of nature when it was pure, untainted, and untouched by the onslaught of foreign invasions. At the end of one article, a resident deplores the dire situation. "I grew up on the backwater," he declares, "and I'm watching it disappear, it's really sad." And the article concludes, "Spoken like a true native" (Verrangia 2000).

ACT II: AND THEN THE TOWERS FELL

Calling 911

On September 11, 2001, the World Trade Center was attacked and the "war on terror" began. This was the first large-scale act of terrorism on the U.S. mainland. The United States (and indeed the world) has endured the specter of 9/11 ever since. In response, the nationalist spirit soared, as did calls for revenge and war. The United States started two wars in Afghanistan and Iraq. On U.S. soil, immigrants, especially those phenotypically "brown" or who looked "Muslim," no longer felt safe as violent acts against immigrants rose. Immigration and law enforcement allied their policies to the larger national and foreign policies. Anti-immigrant rhetoric singed the air as the nation focused on "those who hate America." To be brown was to be an invader, a possible terrorist. And predictably, these sentiments further intensified the earlier images of alien plants and animals (Larson 2007b). Aliens that were once stealthy and silent killers now grew teeth and became more aggressive, menacing, ravenous individuals with extraordinary predatory powers. A case in point is the sensational snakehead (*Channa argus*) that was found in the Crofton Pond in Maryland in June 2002 (Derr and McNamara 2003). Headlines and news stories warned of the land-walking, air-breathing, torpedo-shaped, migrating Chinese predator, with big teeth and a voracious, insatiable appetite, which reproduced rapidly:

> Killer Chinese Fish Surfaces in Maryland (*Action News 6* 2002)
> Stop That Fish! Snakeheads Walk All over Crofton (Ringle 2002)
> Freaky Fish Story Flourishes (Huslin 2002b)
> Crawling Snakehead Fish Scaring Washington (Doggett 2002)
> Freak Fish Found in Two More States (*CBS Evening News* 2002a)
> Invasive "Walking" Fish Found Across U.S. South (Dart 2002)
> Spawn of Snakehead? Freak Fish Have Spawned! Oh My! (Huslin and
>   Ruane 2002)
> Maryland Fears Carnivore Fish Invasion (*Edmonton Journal* 2002)
> A Crusade to Stop the Voracious Fish (O'Brien 2004)
> Chinese Snakehead Not Maryland's Only Foreign Problem (*Daily
>   Record* 2002)

A CNN feature began with asking, "What has a head like a snake, a mouth full of teeth, a long dorsal fin, and the ability to live out of water and waddle around for days at a time?" (*CNN News Online* 2002). The "Asian walkingfish," they charged, was "capable of consuming a pondful of fish and then limping along on its strong pectoral fins and belly to other waters, breathe air and survive for days on land if it stays wet." Some stories claimed they were so highly predatory that they "ate their own young." As an Interior Department spokesperson put it, "These things reproduce quickly. They eat literally anything that's living, including cannibalizing themselves. They'll eat ducklings. They will eat amphibians. So if you leave them in a body of water or give them anything, there'll be nothing left. They're a bad actor" (Huslin 2002c).

Again and again, the headlines invoked the rhetoric of "good" and "evil" and natives as victims, frameworks that resonated with the rhetoric on the war on terror. As the biologist L. B. Slobodkin quipped, "Good species are native species, and oddly enough, the less one might reasonably call them successful, the better they are" (Slobodkin 2001). A CBS story introduced an adult snakehead as a "grown-up killer evil fish" and a juvenile as a "baby killer evil fish." The discovery brought a flood of tourism as people flocked to the region to view the "evil" in the pond. In response, the local capitalist spirit thrived with snakehead paraphernalia. The website snakeheadstuff.com emerged with T-shirts and mugs emblazed with catchy slogans such as "Frankenfish: Marching to a Pond Near You." The robust war on terror spilled over into a "war" with the fish:

Maryland Wages War on Invasive Walking Fish (Mayell 2002)
Wanted Dead: Voracious Walking Fish (*CBS Evening News* 2002b)
Biologists on Mission to Kill (Huslin 2002a)
State declares victory over Snakehead fish (Kobell and Thomson 2002)
Soon Ghastly Fish Will Walk into Sunset (Cowherd 2002)

"Wanted" posters (figure 1) highlighted its "foreign" origins (Derr and Mc-Namara 2003: 127). Campaigns and contests emerged to publicize and exterminate the fish. The Department of Natural Resources launched its own, "Kill a Snakehead to Win"(Department of Natural Resources n.d.). To enter, one had to catch and kill a snakehead (hint: cut the isthmus to kill the fish), photograph the fish with a ruler to show its size, and email the picture for the competition. They were reminiscent of the war on terror and programs like Operation TIPS, which recruited citizens as neighborhood spies. The *CBS Evening News* dramatically aired a story declaring, "No one is sure how many of them are out there, but every one of them is wanted, not dead or alive—just dead" (2002b). Officials proclaimed, "We'll conquer the snakehead. It doesn't matter how big it is, how tough it thinks it is. We'll destroy the snakehead" (Huslin and Ruane 2002). "We are on a 'mission to kill,'" they said (Huslin 2002a). "We want this fish dead. No question about it" (Mayell 2002). One article (Ringle 2002) joked that the paranoid may see the hands of Al Qaeda!

NORTHERN SNAKEHEAD

Distinguishing Features

Long dorsal fin • small head • large mouth • big teeth
length up to 40 inches • weight up to 15 pounds

HAVE YOU SEEN THIS FISH?

[photo of northern snakehead]

The northern snakehead from China is not native to
Maryland waters and could cause serious problems if
introduced into our ecosystem
If you come across this fish,

PLEASE DO NOT RELEASE.

Please KILL this fish by cutting/bleeding
as it can survive out
of water for days and REPORT all catches to
Maryland Department of Natural Resources
Fisheries Service. Thank you

FIGURE 1.

What of the fish's biology? Some news outlets gave voice to these dissensions. Exotic aquatic species is a very old problem (Derr and McNamara 2003). Since the mid-nineteenth century, more than 139 nonindigenous species have been introduced into the Great Lakes and a new species a year into the Chesapeake Bay and the Hudson River system (Mills, Scheurell, et al. 1996). Yet the news stories tended to the sensational. Some biologists argued that such extrapolation tended to the paranoid, "more Hollywood than science" (Kluger

2002). Others argued that the snakehead was nothing more than a "common swamp fish" from Southeast Asia, living in irrigation ditches and rice paddies, thriving until the dry season, before squirming to the next pocket of water. At best its "walk is really more of a wriggle. Such clumsy location does not lend itself to wanderlust, and snakeheads in a good pond are likely to stay forever. Snakeheads are extremely lazy and sedentary," said the Hawaiian biologist Ron Weidenbach (qtd. in Kluger 2002).

Investigating snakeheads as culinary delicacies, vendors said that the best the fish usually do in open markets is several hours under wet burlap and under sunny skies "they are said to fricassee fast" (Kluger 2002). Over the past few years, they have appeared in half a dozen other states but in modest numbers and with limited impact. The snakehead did thrive in Hawaii, where it was aggressively fished (better than bass!—felt many locals). While scientists plotted to exterminate the fish in Maryland, large-scale artificial breeding projects are underway in China and neighboring countries to meet the growing demand for snakehead meat in Asia (Emery 2002).

Where exactly are the Chinese snakeheads from? They arrived in China from India or Southwest Asia during the Pleistocene (Chew and Laubichler 2003). There are fossils of channids in Switzerland and France dating back to the Oligocene (Chew and Hamilton 2011). The history of snakeheads reveals a complex migration history in contrast to the simple origin story from Crofton Pond. As Chew and Hamilton ask, "Do snakeheads *belong* in China? How can a fish demonstrate belonging other than by being, surviving and persisting *here*, *now*, any of which probably exceed its awareness of the issues at hand?" (Chew and Hamilton 2011: 42).

Despite this, the fear of the snakehead rose to fever pitch. Within a couple weeks, the pond was poisoned and all species were killed. The incident was used as an opportunity to publicize the dangers of invasive and exotic species.

> Something from the X-Files: The Frankenfish is here. The invasion has
>   begun. Nothing can stop it (Harrison 2002)
> Juvenile "Frankenfish" Raise the Odds of Alien Invasion (Barnes 2002)
> Invasion of the Snakeheads: Frankenfish is not the scariest import
>   (Fields 2005)

The cycle was revisited the following week after snakeheads were found yet again, this time in Wisconsin. Officials, however, were less worried because of the harsh winters.

ACT III: THE OBLIGATION OF RELUCTANCE

In 2006, an interesting op-ed entitled "Border War" appeared in the editorial spread of *The New York Times* (Ball 2006). It began:

> The horticultural world is having its own debate over immigration, with some environmentalists warning about the dangers of so-called exotic plants from other countries and continents "invading" American gardens. These botanical xenophobes say that a pristine natural state exists in our yards and that to disturb it is both sinful and calamitous. In their view, exotic plants will swallow your garden, your neighbors' gardens and your neighbors' neighbors' gardens until the ecosystem collapses under their rampant suffocating growth.

I was delighted! The author went on:

> If anything suffocates us, though, it will be the environmentalists' narrow mindedness. Like all utopian visions, their dream beckons us into a perfect and rational natural world where nothing ever changes—a world that never existed and never will.

My heart sang! Then the author concluded:

> Let's welcome, as spring arrives tomorrow, as many huddled masses of flowers, herbs and vegetables as can fit in our unique melting pot of a nation, unrivaled in its tradition of lush diversity and freedom to grow rampantly.

This was music to my ears. In the center pages of *The New York Times,* someone acknowledged the dense traffic of meanings between our worlds of natures and cultures, of the ways in which we transfer our cultural anxieties into law and order of the plants and animals that house our gardens and streets. Having worked on this issue for many years, I welcomed the author's sentiments and politics. Delighted, I filed it away. Until one of my colleagues pointed out who the author was—George Ball, president of the Atlee, Burpee and Co, a seed company. I came to a halt and had to ponder what had happened. This op-ed appeared during a resurgence of the immigration debate within the United States. We were witness to the rabid xenophobic rants of Congressman James Sensenbrenner and his House Bill 4437, the Border Protection, Antiterrorism and Illegal Immigration Control Act of 2005, which charged any undocumented immigrant and anyone who "assisted" her or him with "aggravated felony." We have seen the rise of the Minutemen vigilantes and the growing number of deaths at the border due to the militarization of Operation Gatekeeper (Coronado 2006). But we also saw more than one million people march to protest the anti-immigration legislation in LA and across the country. Labor activists of yesteryear reappeared in the papers. People stayed away from work in support. And we saw a debate within the Republican Party as President Bush, Senator McCain, and others attempted to carve a path that moderated the criminalizing of immigration in the Sensenbrenner position. And yet others on the right advocated an open door to immigration. March was an important month when these debates unfolded, and perhaps it was no surprise that the head of a seed company who is in the business of trying to sell seeds of all varieties (and not only native ones) should favor an open policy of unbridled choice for our gardens.

In another interview, Ball goes on to further elaborate. Weeds, he argues,

are not considered desirable. Weeds are invasive, uncontrollable plants ... Taboo plants. A lot of weeds are so called because they take off in all directions, they have qualities of sinfulness. You know this is a garden. There are certain things you just simply can't have in your yard if you want to have these other objects, and that's the border between desirable and undesirable. God and the Devil ... Here are these little tiny white flowers with shabby little green leaves that this plant has worked with all its strength, its ancient lineage of energy, of genetic reality, to produce, for whatever reason. Who am I to pull that out? And there I was, looking at it in my hand. In what way was this better or worse than the lily of the valley? I felt so stricken by my own complicity.

Here in hyperbole with allusions to hard work, beauty, and ancient pedigree was the righteous and ethical angst of the recently native. And yet these were fueled by the politics of an ebullience of commerce, of unbridled choice in our economies and gardens. I favored one but not the other. It has been a difficult lesson to learn in our times. A strange confluence of bedfellows as notions of purity and authenticity haunt our imaginations of science and the natural. We see this rhetoric not only in our discourses on invasive species, but also on genetically modified organisms and the new reproductive technologies. Rhetoric is never innocent. Histories of gender, race, and class politics are central to why certain ideas resonate and take hold. It would seem that the political right and left have both inherited and indeed embraced the colonial imaginary.

As George Ball argues in an interview, "I find extraordinary the whole question of hybrids and the sentimental fondness for older plants because I find that sentiment in all of our attitudes about philosophy, religion, et cetera—that old-time religion, that fundamentalist religion. When people stop being religious fundamentalists they switch over to being food purists; it simply transposes itself from one domain to another." I cannot but celebrate such sentiments when contrasted with the often xenophobic and nativist politics and sentiments of many conservationist groups that I otherwise support. And yet while the inclusivity of George Ball resonates, I am troubled with the voracious appetite of global capitalism that will sell anything—even at the cost of habitat destruction, the debt and dependence of family farms, and the embrace of high-input agriculture. Complex times call for complex measures. I feel the obligation of reluctance—the obligation to no ideology where all means justify the end and a deep reluctance to agree with any position without a thorough investigation of the political, ecological, historical, and economic roots as well as consequences of those positions.

ACT IV: THE INVASIVE SPECIES WAR

In April 2011, Hugh Raffles published an op-ed in *The New York Times* entitled "Mother Nature's Melting Pot." Two months later, in a June 2011 issue of *Nature*, M. A. Davis and eighteen other biologists wrote a commentary cautioning, "Don't judge species on their origins" (Davis et al. 2011a). There were high-profile responses to and comments on both pieces. Several articles in *Science* also responded to the Davis et al. piece. In July 2011, Leon Neyfakh reported on this public turn in the *Boston Globe* as "The invasive species war." All three well-publicized pieces were making similar points, many of which I have already outlined earlier in this chapter. In spirit, all argue that "'non-native' species have been unfairly vilified for driving 'beloved native' species to extinction, and polluting 'natural' environments,'" thereby creating "a pervasive bias" against alien species (M. A. Davis et al. 2011a: 153). After all, not all native species are well adapted to local communities, nor all foreign species maladapted to them. The biological world presents too much heterogeneity and complexity for such simplistic generalizations (Hettinger 2011). After all, the increase of invasive species has happened alongside climate change and other land use changes. Indeed, given the dynamic nature of evolution, some natives may well be destined to be temporary residents in the grand scheme of evolutionary time—with or without humans (Raffles 2011). We should abandon such thinking, these pieces argued, since the native/alien binary while still a "core guiding principle" in biology was declining in its usefulness, even becoming counterproductive.

Shifts in species composition have been ubiquitous in evolutionary history and should not surprise us. Despite many "apocalyptic" scenarios of invasions, major extinction threats are not backed by data (Richardson 2011). Most campaigns to eradicate invasive species simply have not worked, these articles argue.

In contrast, they remind us that new arrivals can often help an ecosystem rather than only hurt it; alien species have often increased biodiversity and helped habitats, allowing native insects and birds to flourish (Davis et al. 2011a, Neyfakh 2011). Many of our prized flora and fauna, such as honeybees, are foreign in origin, yet economically invaluable (Raffles 2011). The quintessential day lilies and Queen Anne's lace that beautify the New England roadsides are in fact aliens (Pollan 1994). For example, the ring necked pheasant, state bird of South Dakota, the purple lilac, state flower of New Hampshire, and the red clover of Vermont are all foreign in origin (M. A. Davis et al. 2011a, Neyfakh 2011). In contrast, some native species such as the pine beetle (*Dendroctonus ponderosae*) and many species of native barnacles have proven to be invasive and have caused great damage (Hettinger 2011). Others go further in arguing that invasives in fact restore Earth's ecosystems—growing where nothing else will, they remove toxins and restore health to habitats (Timothy Lee Scott and Buhmer 2010).

All these pieces conclude that the selection of any historical period as "native" is entirely arbitrary. Classifying organisms by their "adherence to cultural standards of belonging, citizenship, fair play and morality does not advance our understanding of ecology" (M. A. Davis et al. 2011a: 154). Instead, we ought to embrace a more dynamic and pragmatic approach, focusing on the function of species in their ecosystem rather than making their geography of origin a litmus test. Raffles connects the anti-immigrant crusades of the Minutemen and the Tea Party as well as the native species movement led by environmentalists, conservationists, and gardeners to a widespread fear captured in Margaret Thatcher's well-known phrase of being "swamped by aliens." Instead, plants and animals, like humans, need a "thoughtful and inclusive response" (Raffles 2011).

Neyfakh begins his piece with the Charles River Watershed Association's weekend campaigns to eradicate the European water chestnut. The "No More Water Chestnuts" campaign, he argues, can attract up to seventy volunteers on a Saturday morning. Volunteers pull out the undesired plants by their roots and collect them in plastic laundry baskets. Similarly, towns have begun to regulate the number of alien plants in home gardens. What breeds such passion? Here the responses to the above pieces are revealing. In response to Davis and his 18 colleagues, Daniel Simberloff along with 141 colleagues argue that the original piece was the "slander of ideologically driven contrarians" rather than scientific wisdom (Simberloff et al. 2011). These biologists argue that Davis et al. are creating two "straw men." First, they refute that most conservation biologists and ecologists oppose non-native species per se. Second, that they ignore the benefits of introduced species. Other letters by Lerdau and Wikham,

Alyokhin and Lockwood et al. all concur and vehemently disagree with Davis et al. They argue that there is no campaign against all introductions or the eradication of all introduced species. Finally, they accuse Davis et al. of downplaying the severity of the impact of non-native species, especially those that do not manifest their invasiveness until decades after their introduction (Lockwood, Hoopes, and Marchetti 2011). Further, while indeed new entrants may appear to increase local biodiversity, the loss of local species decreases global diversity (Alyokhin 2011). Invasive species policies have seen considerable success in eradicating invasives (Lockwood 2011). Lerdau and Wikham caution that waiting and watching is often unsuccessful; it is cheapest, easier, and most effective to eradicate foreign species soon after detection (Lerdau and Wikham 2011, Haack et al. 2010). They insist that harmful species must be kept in check using biological, chemical, or mechanical means (Simberloff et al. 2011). Simberloff et al. (2011) conclude that the ills of alien and exotic species is a "non-debate" in the scientific community, lest politicians use the essay to cut environmental funding for the eradication of invasive species.

In reading the original commentaries and the voluminous responses and comments, on the surface, it seems that both sides aren't so far apart. Both sides agree that only a fraction of the non-native species are harmful; when they are, all agree that they can be destructive and need to be reined in. But how do you tell which "alien" will become the "invasive"? Nip it in the bud, some biologists say. Catch it before it becomes a problem! Similarly, how do you tell which "alien" will become the "terrorist"? Again, our immigration policy has been about nipping it in the bud—"racially'" profiling particular nations, whether they are citizens or not.

Perhaps it should not surprise us that all three pieces appeared in the summer of 2011. The pro-immigrant sentiments were in the air. The Tea Party was at the helm in Congress, holding the nation hostage in the debt crisis talks. Anti-immigrant, xenophobic vitriol pervaded the country as several states passed anti-immigrant laws that targeted anyone looking "foreign." In response, the pro-immigration activists launched a national and visible debate about the dangers and benefits of immigration. California came to be the third state to pass a version of the Dream Act, which allows the undocumented to receive government tuition aid. Thirteen states allow undocumented students to qualify for in-state tuition rates (*New York Times* Editorial 2012). In a political climate where immigration is hotly debated, pro-immigration and pro-invasive rhetoric should not surprise us. The invasive species "war" reflects our continuing ambivalence about immigration.

The Rhetoric of Biological Invasions

In this debate, my sympathies lie with Davis et al. There are striking similarities in the qualities ascribed to foreign plants, animals, and people, and these debates track each other. The xenophobic rhetoric is unmistakable. The point of my analysis is not to suggest that we are not losing native species, or that we should allow the free flow of plants and animals in the name of modernity or globalization. Instead, it is to suggest that we are living in a cultural moment where the anxieties of globalization are feeding nationalisms through xenophobia. The battle against exotic and alien plants is a symptom of a campaign that misplaces and displaces anxieties about economic, social, political, and cultural changes onto outsiders and foreigners.

The obsession with native/alien reflects something deeper—a pervasive nativism in environmentalism and conservation biology that makes environmentalists biased against alien species (Pollan 1994, Paretti 1998, Warren 2007). For example, the final chapter of one of the many recent books on the topic is entitled "Going Local: Personal Actions for a Native Planet." Such rhetoric conjures up a vision where everything is in its "rightful" place in the world and where everyone is a "native" (Van Driesche and Van Driesche 2000). In this debate, the "natives" are ironically the white settlers, not the original natives.

I want to be clear that I am not without sympathy or concern about the destruction of habitats, which is indeed alarming. In their zeal to draw attention to the loss of habitats, however, some journalists and scientists feed on the xenophobia rampant in a changing world. They focus less on the degradation of habitats and more on alien/exotic plants and animals as the main and even sole problem. In contrast to humans, where the politics of class and race is essential, the language of biological invasions renders *all* outsiders (Devine 1999). Conservation of habitats and our flora and fauna need not come at the expense of immigrants.

Instead, let us consider exotic/alien species in their diversities. Mark Sagoff points out that the broad generalizations of exotic/alien plants obscure the heterogeneity of the life histories, ecologies, and contributions of native and exotic plants (Sagoff 2000, 2005). For example, he points out that nearly all U.S. crops are exotic plants while most of the insects that cause crop damage are native species. Ironically, alongside a campaign against foreign species, there is simultaneously another campaign that promotes the widespread use of technologically bred, genetically modified organisms for agricultural purposes.[4] In these cases, the ecological dangers of growing genetically modified crops in large fields are presented as minimal. Concerns of cross-fertilization with native and wild plants are dismissed as antiscience/antitechnology. Ultimately, it would seem that it is a matter of control, discipline, and capital. As long as exotic/alien plants know their rightful place as workers, laborers, and providers, and controlled commodities, their positions manipulated and controlled by the natives, their presence is tolerated. Once they are accused of unruly practices that prevent them from staying in their subservient place, they threaten the natural order of things.[5]

What is most disturbing about displacing anxieties attending contemporary politics onto alien/exotic plants is that other potential loci of problems are obscured. For example, some scholars point to the fact that exotic/alien plants are most often found on disturbed sites. Panic about aliens minimizes significant destruction of natural habitats through overdevelopment. A displacement of the problem onto the intrinsic "qualities" of exotic/alien plants and not onto their degraded habitats produces misguided management policies. Rather than preserving land and checking development, we instead put resources into policing boundaries and borders and blaming foreign and alien plants for an ever-increasing problem. Unchecked development, weak environmental controls, and the free flow of plants and animals across nations all serve certain economic interests in contemporary globalization. Ultimately, the campaign against the foreign does not solve species extinctions or habitat degradation.

The Ghosts of Naturecultures

Rhetoric, words, and symbols play a central role in the campaigns against invasive species because anti-immigrant rhetoric latches onto long-enduring, powerful, and familiar tropes of fear (B. Hartmann et al. 2005). The fear of the outsider, impurity, and pollution tracks centuries of campaigns that have rendered some bodies as violent, frightening, undesirable, and worthy of being controlled, even exterminated. The fears about cross-fertilization of foreign and native plants are eerily reminiscent of fears and laws against human miscegenation. The ideologies that produced this history simmer beneath the anti-immigrant rhetoric that helps contemporary campaigns rouse strong sentiments and mobilization.

What is the history that simmers beneath? It is the lives of the ghostly dispossessed—the bodies displaced, starved, colonized, violated, sterilized, experimented upon, maimed, killed, exterminated. In the realm of nature, there is increasing attention to the destruction of forests, conservation, preservation of native forests and lands, the commodification of organisms, and concern over the invasion and destruction of native habitats through alien plant and animal invasions. Yet how does one launch a campaign to save the forests? The most successful have been to draw on tropes of fear and terror. Like the successes of the "population bomb" that engendered visions of the brown hordes at our borders, invasive species rhetoric draws on the same tropes of a beleaguered white nation under siege by the violent and invasive hordes of the brown, black, and yellow. A naturecultural lens allows us to see the underlying nativism—an ideology where nature and people are "in place."

Invasion ecologists and environmentalists are often embattled in making a case for the environment. The fact that climate change is even debated is a case in point. Environmentalists, often members of the progressive left and supporters of peaceful, inclusive visions of the world, recoil at being labeled "conservative, backwards—even intolerant" (Neyhfakh 2011). But in order to bring attention to the crisis of the environment, must we resort to age-old xenophobic and racist attacks? A naturecultural lens allows us to see the immense destructive power of such campaigns and what is lost when xenophobic, racist, and sexist sentiments undergird a worthy goal. Whom are we saving and what is lost in such saving?

Indeed, while Davis et al. and their critics may agree that only some species are destructive, and while officials agree that only a few individuals resort to terrorism, the deeper political and philosophical question is, what of the "other" others? In the very act of labeling humanity and biota into two categories—native and alien—marking the presumed good from the possible evil, our quest for an inclusive, ethical world is lost. There are other characteristics—biological and social—that are useful; foreignness need not be the sole criteria (N. Davis 2009). After all, the most invasive species is the human, especially the elite, white, western human!

To what lengths will we go to "restore" our world to some nostalgic vision of the past? Whose nostalgia? While invasive species do damage, so do roads and "green" bioenergy plants that are erected in service of our communities. Restoration ecology as a field has embraced biological, mechanical, and chemical interventions with gusto. Small orange flags mark the sites of our increasingly herbicide-ridden landscapes. Will we chemically bombard ourselves to our nostalgia? What does it mean to "restore" our world, while we pollute our soil and ground water? What are we saving, and for whom?

Feminist and postcolonial critics of science have shown us repeatedly how political, economic, and cultural factors inform and shape scientific questions, answers, practices, and rhetoric. Both the cultural and scientific worlds house diverse and heterogeneous views with a long tradition of dissent. Many ecologists and conservation biologists have developed alternate models, challenging the dominant framework of conservation biology (D. Keller and Golley 2000).

Studying "naturecultures" means being cognizant of how science and the humanities are embedded in naturecultural contexts. Just as science does not mirror nature, we must not reduce science to mirroring politics either—right or left.

Living in Naturecultures

Susan Matt reminds us that in the nineteenth century, millions of immigrants who were homesick from the emotional toll of migration were diagnosed by a medical condition with its clinical name: nostalgia (Matt 2012). Many generations later, in the twenty-first century, we are suffering the same nostalgia! What does it mean to be homesick for a world that is two centuries old, a world we have never seen but in our nostalgic visions?

This is a story in flux. It is not clear if U.S. politics and society will rework its rhetoric against aliens (flora, fauna, or humans) or if anti-invasive rhetoric imbued with a simplistic anti-immigrant rhetoric will prevail. If history is any indication, this is not a linear story and we are not at the story's end. Society will revisit, reshape, and rethink its ideas of national belonging and of natives and aliens for centuries to come. As long as we do not resolve the fundamental question—about variation, diversity, and difference—these debates will continue to rage in science and society, in defining our natural and cultural worlds. And to be sure, each time these debates are renewed, we will debate them as new problems that we have never encountered before!

CHAPTER FIVE

My Experiments with Truth

Studying the Biology of Invasions

The truth will set you free, but first it will piss you off.
—Gloria Steinem when accepting an award, 2005

I have the nerve to walk my own way, however hard,
in my search for reality, rather than climb upon the
rattling wagon of wishful illusions.
—Zora Neale Hurston, "Letter to Countee Cullen"

Over the past three decades feminist scholars have amply demonstrated that critical social categories such as gender, race, class, sexuality, and nation are inextricably interconnected with science and its studies of nature. Science has played a central role in shaping these categories, and these categories have in turn shaped science. For the most part, this endeavor has focused on historical works. If social and political factors have always been important to science in the past, surely they must also be to the science of the present? If the history of science teaches us that science has always been socially embedded, surely science continues to be socially embedded? How does one practice science with that knowledge? That is, how do we study the biology of naturecultures?

As we saw in the previous chapter, invasion biology as a field can be understood to be deeply embedded in its political and cultural times. We have also seen that this is not a monolithic field. The scientific world continues to debate the terms and interpretations of native and foreign species and what it means for our environment (Hattingh 2011, Larson 2007a). What would it mean to engage the feminist studies of science *in* the practice of science? This is the project that has engaged me—to use the historical, sociological, rhetorical, and philosophical insights of the cultural and feminist studies of science *in* the practices of experimental biology. A project that brought the sciences and the social studies of science, the humanities and the biological sciences, into conversation with each other in order to together produce knowledge about the natural world. I believed I found the perfect project in exploring the biology of plant invasions. This chapter presents one example of how we can experiment with an interdisciplinary repertoire of research questions, methods, and epistemologies to produce knowledge about the biological world—an experiment about experimenting!

The experiments in this section focus on biological invasions. We have seen in the previous chapter that in recent years, there has been considerable panic and hype about invasive plant species. The alarm is framed around the "identity" or foreign origins of plant species rather than about the ecological or environmental contexts of invasion, that is, a greater focus on the invasive plant rather than on the destruction of local ecologies. The response in turn has been about identifying and eradicating the troubling species. For example, local environmental societies lead their communities in weekend missions to pull out noxious species. Popular ideas of plant species thus tend to personify organisms into sensationalized notions of good native and evil foreign species. While less sensational and categorical, the same ideas permeate scientific work and the views of some scientists.

The experiments on the biology of invasions progressed alongside my growing understanding of the increasing number of critiques and alternate visions of invasion biology that have emerged over the past few decades (J. H. Brown and Sax 2004, 2005, Calautti and MacIsaac 2004, Chew and Hamilton 2011, Coates 2003, 2006, M. A. Davis 2009, M. A. Davis et al. 2011a, Gobster 2005, Hattingh 2011, Larson 2007c, Sagoff 2000, 2005, Slobodkin 2001, Subramaniam 2001, Theodoropoulos 2003, Townsend 2005, Warren 2007). While some biologists have continued in the tradition of viewing humans as outside of nature, others have worked hard to incorporate humans into the natural world and to bridge the worlds of science and society (Bradshaw and Bekoff 2001, Dietz and Stern 1998, Odum 1997, De Laplante 2004). Despite the strident tones of many biologists, even Charles Elton, often regarded as the father of invasion biology, was rather moderate in his views.

> I believe that conservation should mean the keeping or putting in the landscape of the greatest possible ecological variety—in the world, in every continent or island, and so far as is practicable in every district. And provided the native species have their place, I see no reason why the reconstitution of

communities to make them rich and interesting and stable should not include a careful selection of exotic forms, especially as many of these are in any case going to arrive in due course and occupy some niche. (Elton 1958: 155)

Finding a way to think natureculturally was particularly important for the experiments we were undertaking. In Southern California where the project was based, human-made disturbance has a very long and destructive history. Ranching and cattle grazing over hundreds of years have considerably altered the landscape. In a state with long histories of biotic migration, questions of what are deemed native and exotic are particularly and deeply fraught. How do we move beyond ideas of a picture-perfect "nature," which we artificially maintain? How do we understand the human species as part of nature, in all its shifts and evolutions? I came to understand these as important questions that can guide biologists in the development of experimental research. Is it possible to characterize exotic/native plants? Do they all share common life history parameters and ecological traits? How heterogeneous and diverse are the species within those categories? How static and co-evolved are native communities? What is the relationship of plants and their soil communities and what impact do exotic plants have on them? Do they destroy and degrade these communities? As ecologists, we can test these theories, intervene, and participate in the national conversation not only on exotic plants, but also on immigration and race relations.

As we have seen, the binary formulation of nature/culture has been thoroughly critiqued (Demerritt 1998, Valentine 2004). Our definitions of what is natural/unnatural, pristine/degraded, or authentic/fake remain arbitrary, ambiguous, contested, and fuzzy and make poor grounds for biological and environmental policy. Yet there remains tremendous power in the binary tropes that shape environmental thinking; despite the critiques, the binary black/white vision endures and "sparks our passions but darkens our vision" (Cronon 1996b: 39). The real problem is that such a vision ultimately leads us "back to the wrong nature" (Cronon 1996a: 69).

More important, these formulations have shaped not only our language but also the theories and discipline of ecology. When the field of invasion biology emerged in the 1980s through SCOPE (Scientific Committee on Problems of the Environment), most of the biologists were ecologists, particularly community ecologists (M. A. Davis 2011b). As a result, invasion biology emerged as a disciplinary offspring of community ecology and is still dominated by the niche-based theories of MacArthur and Hutchinson. In its early years, deterministic models were emphasized, focusing on local processes. For example, the connection of diversity and invasibility remains a strong idea—species-rich environments are more resistant to invasion than species-poor ones. Like Elton and Darwin's naturalization hypothesis, the theories tend to focus on species traits and local and deterministic factors. Even while empirical studies contradict the easy connection between species richness and resistance to invasion, niche-based theories continue to be overused at the expense of regional and historical factors. Invasion biology has seen much more theory generation than rejection, and as a result theories explaining invasions have tended to accumulate. Many feel that these factors have impeded the growth of the field (M. A. Davis 2011b). Indeed, "those who believe their policy towards aliens is determined by some objective standards are living in ivory towers" (Mabey 2005: 46).

Some scholars question the disjuncture being the empirical evidence of ecology when compared to its theoretical claims. For example, examining terms such as *community* or *ecosystem* shows them to be rather murky and changing over time (Sagoff 2006). Despite the vast literature in ecology, there is a "stunning lack of evidence that natural communities or ecosystems possess any mode of organization that can serve as the object of 'environmental protection'" (Sagoff 2006: 157). In fact, there is little consensus on what constitutes an ecosystem or how natural selection structures ecosystems. Rather than a science with consensus, the empirical evidence suggests that ecosystems and communities lack organization, structure, or function and are rather temporary, contingent, and even ephemeral accidents of history (Sagoff 2006). Indeed, the idiosyncrasies of ecologies of communities and ecosystems could be a law in itself. Normative and intrinsic values of concepts like ecosystems and biodiversity, especially in a world with profound human intervention, remain problematic (Hattingh 2011). This line of reasoning—of unraveling the social, cultural, and ideological assumptions of biological concepts—is powerful, and many biologists have raised them. These issues are prevalent in invasion biology. Stephen Jay Gould summarizes this well in arguing that native plants are

> only those organisms that first happened to gain and keep a footing . . . In this context, the only conceivable rationale for the moral or practical superiority of "natives" (read first-comers) must lie in a romanticized notion that old inhabitants learn to live in ecological harmony with surroundings, while later interlopers tend to be exploiters. But this notion, however popular among "new agers," must be dismissed as romantic drivel. (Gould 1997: 17)

With respect to invasive species, many biologists question whether the hype and vilification of foreign species is warranted, given the empirical evidence except perhaps in the case of small islands. Yet it is repeatedly touted that invasive species are the second greatest threat for the extinction of species (P. Roberts et al. 2013, D. H. Richardson 2011, M. A. Davis 2009). Invasive species have received far greater press than many other environmental problems. Indeed, dealing with human-induced environmental effects and the impact of climate change is more urgent than ever and likely to intersect with lots of fields in biology and outside (D. M. Richardson 2011).

This intellectual context with a history of reflexive critiques on language, theories, and philosophies of ecology seemed an ideal context in which to begin to think natureculturally. The project on biological invasions was a collaborative project with two biologists, James Bever and Peggy Schultz.[1] Jim and Peggy's research interests involve arbuscular mycorrhizal (AM) fungi and their role in plant ecology. Arbuscular mycorrhizae are characterized by the arbuscules, or vesicles, they form with the roots of vascular plants. Belonging to the phylum *Glomeromycota*, these fungi develop a symbiotic association with the roots of vascular plants. The fungi help plants capture nutrients such as phosphorus, sulfur, nitrogen, and other micronutrients from the soil and in turn the plants house the fungi in their roots. In addition, AM fungi also physically affect the soil through the production of glomalin, a sticky substance that binds soil particles and improves soil aggregation (Wright and Upadhyaya 1998, Chaudhary and Griswold 2001). It is believed that more than 80 percent of vascular plants species form mycorrhizal associations (Harley and Smith 1983). While sometimes mildly pathogenic, mycorrhizal fungi are usually in highly evolved mutually beneficial or mutualistic associations with plants. AM fungi are believed to have played a critical role in the evolution of vascular plants and their colonization of land (Brundrett 2002). Jim and Peggy's lab is broadly interested in the biology and ecology of these fungi, and they proved to be great colleagues and collaborators. Politically engaged, they were supportive of my research goals. What excited me about their work and approach to invasive species was moving beyond the usual "geographic origin" story approach to invasion ecology; instead, they took seriously the ecological contexts of plants. In paying attention to the environmental contexts of growth, especially the soil communities of plants, their work held exciting potential for collaboration. Mycorrhizal fungi and their relationship with native and exotic plant species seemed like a great context for a science/science studies project. Their work on fungi that were in "mutualistic" relationships also challenged the role of competition as the critical driver of ecology and evolution of plants. Living in Southern California where human-made landscape changes are considerable, they were interested in invasion biology themselves, and thus this project on the role of mycorrhizal fungi on invasion biology emerged as a collaborative one. I was particularly mindful of the intensity of time that fieldwork and subsequent data gathering would take during busy semesters when my job involved teaching in a humanities department. It was a daunting task to fit in with the challenges of interdisciplinary work. A collaborative project with supportive colleagues who had an established laboratory seemed an ideal context in which to explore these questions, and indeed it was.

Studying Invasion Biology

How can you do interdisciplinary work across disciplines that have no channels of official communication? Geographically separate on campuses with no joint conferences, no journals, no interdisciplinary committees, and no hallway conversations? As we saw in the previous chapter, some biologists have long been critical and reflective of the field of invasion biology (Hobbs and Richardson 2011, Preston 2009, D. M. Richardson and Pysek 2008, Simberloff 2003, Soulé and Lease 1995, Vermeij 2005), and yet the field has continued on rather unaffected. The strategy was to do interdisciplinary work with multidisciplinary audiences in mind. This way the work could be published in several disciplinary journals with some of the interdisciplinary insights being translated into the language of disciplines. The interdisciplinary methodology could be published in interdisciplinary sites. Given the institutional structures, it was the only way I could conceive of funding and carrying out such work. As will become evident later in the chapter, one of the tautological challenges of invasion biology is that testing ecological theories necessitated working with the categories of native and naturalized.

RESEARCH QUESTIONS

The experiments we carried out were designed to examine the relationship between native and naturalized species and their soil communities, in particular, AM fungi. Did native species perform better than naturalized species when grown with native soil communities? Were naturalized species less dependent on soil communities than native species? Did AM fungi grow better with native species? The experiments thus tested whether the ecological context of plants with respect to their soil communities mattered to the success of native and naturalized species. Similarly, did the composition of soil communities shift depending on the plants that grew in them? Can we really talk about the broad categories of native and naturalized as though they were homogeneous? Do all native and naturalized species act in identical ways whereby these categories of native and alien/naturalized make biological sense?

EXPERIMENTAL DESIGN AND METHODS

In this experiment we worked with a range of native and naturalized plant species to see if the categories of native and naturalized mattered. First, we did a floral survey of plant families in Southern California to evaluate whether native and naturalized species varied in their dependence on AM fungi. We also developed cultures of soils drawn from native and naturalized species to see if the density of AM fungi in the soil varied depending on which plant species grew with them.

The field experiments were conducted in Southern California, and all the seeds and inoculum for the soil were collected from UCI Ecological Preserve and the Starr Ranch Wildlife Sanctuary in Orange County. We tried to maximize the range of families represented in our experiment to include species that were known to associate with AM fungi as well as those that were not.

Seeds were collected and then propagated in the greenhouse. They were randomly assigned to the various treatments to ensure uniformity. Just as seeds were propagated, soil organisms were propagated by collected soil from the same locations as the seeds and then grown in the greenhouse to create soil inocula that were used to develop native and exotic soil communities. In order to create the inoculum for native and exotic soil communities, we collected soil from areas that were dominated by natives and those dominated by exotic/naturalized species; these became the "native" soil and "exotic" soil treatments in the experiments. The soil treatments were then mixed with an autoclaved (sterilized) mixture of local soil and sand and grown with nine different native grassland species. All the soils derived from native and exotic sites were pooled to create the experimental treatments of native and exotic soil. For sterile treatments, the same sources were mixed but autoclaved so the soil was sterile. Tests confirmed that the soils were indeed sterile.

We worked with 14 species, 6 native and 8 naturalized. The experiment was conducted over 2 stages.[2] In the first stage, each of the 14 native and naturalized species was grown in 3 soil treatments: native, naturalized, and sterile. Did native plants do better in native soil than exotic or sterile soil? Did naturalized species do better in native soil, or did they prefer exotic soils? Each of the 14 species grown in 3 treatments was replicated 5 times ($14 \times 3 \times 5 = 210$). The 210 pots were randomly placed in the field, and the growth of the plants was measured each month. At the end of the experiment, the seeds and plant and root biomass of the species was measured.

While for stage one, we had broad categories of inoculum—native, exotic, and sterile, at the end of stage one, we had specific soil for each of the 14 species. For stage two of the experiment we used plant species-specific soils as inoculum and grew each plant species in the soil of every other plant species including its own—each combination was replicated 3 times. Plants were also grown in their own cultivated soil, and these were replicated 10 times. So we were able to test not only how native species and exotic species collectively performed in native, exotic, and sterile soils, but also how native species grew in the individual soils of other native and naturalized plant species. Did native species do best in their own soils? Did they do equally well in all native soils? What about exotic species? The 636 plants were randomly divided into 3 spatial blocks and the growth of the plants was measured each month. At the end of the experiment, the seeds and plant and root biomass of the species were measured to evaluate the growth success of the plants.

RESULTS

Several significant pieces of evidence emerged that are worth highlighting here.

FLORAL SURVEY

Southern California is a site of tremendous anthropogenic, or human-induced, change. Cattle ranching over hundreds of years has profoundly affected the landscape of Southern California. The floral survey showed that while all sites had native and naturalized species, a greater proportion of naturalized species in these sites belong to nonmycorrhizal families (i.e., plant families that did not associate with mutualistic mycorrhizal fungi). This was true in all the locations we examined except one, San Mateo Canyon Wilderness area. Strikingly, San Mateo Canyon Wilderness area has remained largely undisturbed and also has the lowest percentage of nonmycorrhizal plant species. This reinforces the idea that disturbance is important to the success of nonmycorrhizal species and that reduced mycorrhizal dependence of naturalized species may be a source for the success of naturalized species (Murray Frank, and Gehring 2010). The pattern shows that species with weak associations with mycorrhizal fungi also are highest in areas with anthropogenic change, emphasizing the critical role of disturbance in changing species compositions. As a result, mycorrhizal naturalized species are underrepresented in our study, a factor that will come to shape its results.

EXPERIMENTAL RESULTS

Overall, the results reinforced many of the patterns predicted in the literature. Native plant species had higher mycorrhizal infection rates than naturalized species, and their growth rates were higher in native soil than in exotic soil. Naturalized species were less dependent and responsive to mycorrhizal fungi than native species (even among species that did associate with mychorrhizal fungi). Naturalized species, on the other hand, did not show a significant response to the soil treatments, that is, they did not do substantially better in one soil treatment (native, exotic, or sterile) compared to others.

To make sure that we were not making false assumptions about the soil, we tested the inoculum densities in native- and exotic-dominated sites and indeed there was a significantly higher density of mychorrhizal fungi in native-dominated soil than in exotic-dominated soil.

Indeed, if you pool the data and look at the categories of native and naturalized, natives overall do better in native soil and naturalized species show a greater adaptability and ability to grow in a range of soil types. But using the broad categories of native and naturalized also obscures the heterogeneity inherent in the diversity of species and range of life histories of natives and naturalized species. While the categories of native and naturalized show the broad patterns, in each of the experiments there was significant variation among native and naturalized species. While some natives did better in native soil than exotic soil, there was variation in that some did much better than others. Some naturalized species did better in native soil than exotic soil and some did better in sterile soil than either native or exotic soil. Thus the heterogeneity is striking and worth noting. One of the other significant aspects of the results was ascertaining feedback cycles. In a positive feedback cycle, plants do well in their soil communities, which in turn do better with those plants. Thus we can see a positive feedback cycle where plant communities and soil communities are caught in a cycle that facilitates the growth and the success of each other. But plants can also show negative feedback, and here plant communities may hinder the growth of soil communities, or vice versa. In the experiments, naturalized species show greater negative feedback, than native plants, that is, they did better in the soil of other naturalized species than in their own soil. While the pattern was overall more marked in naturalized species, some native species also show negative feedback. Disturbance of soil particularly aids plant species in negative feedback loops.

THE COMPLEXITIES OF THE RESULTS: A DISCUSSION

What is in part being lost in the broad categories of natives and naturalized is the variation and heterogeneity within each of the categories. There is considerable variation among native and naturalized plant species. In particular, response to mycorrhizal fungi varies considerably across naturalized species, and this may be one of the multiple reasons for their success. Ecologists suggest that certain life history characteristics enable plants to thrive in disturbed habitats. Traits such as high fecundity, rapid growth, self-compatible breeding systems, and efficient dispersal mechanisms are all traits that have been shown to be crucial to success. Not all naturalized species become invasive, and some native plants can also be invasive. Association with mycorrhizal fungi has been suggested as yet another variable that shapes plant and soil communities (Pringle et al. 2009). Furthermore, naturalized species do not do well in all contexts—they appear to thrive in habitats with low species diversity, high heterogeneity, and, most important, disturbance. Naturalized species also appear to have fewer pathogens in their new environments.

The results suggest that we *cannot* understand plant invasions without considering the ecological contexts of plants. The success of some species may occur through the central role soil communities, especially mycorrhizal fungi, play (Stampe and Daehler 2003). The complex feedback loops that emerge between individual plant species and their preference for individual AM fungal species can fundamentally alter the species compositions of both plant and soil communities and ultimately the future composition of both plant and soil communities (Batten et al. 2008, Marler et al. 1999, Zhang et al. 2010). Some naturalized species do particularly well in disturbed habitats. Let us not forget that disturbance also alters species composition among native species and that native species can also be invasive. But the species that thrive in disturbed habitats tend to be non-AM fungi dependent and thrive in disturbed habitats. Nonmycorrhizal plants erode soil communities, which in turn makes it difficult for native species to reestablish.

The experiments also show that weaker dependence of naturalized species on mycorrhizal fungi affected soil communities of fungi. Overall, naturalized species sustained lower AM fungi densities. There was variability among naturalized species' response to AM fungi, however. Some naturalized species grew better with AM fungi but were poor hosts to the AM fungi, while others did poorly in the presence of AM fungi but proved to be excellent hosts. Therefore, the patterns of establishment and reestablishment of native species depended a great deal on the exact species composition of the natives and naturalized.

Disturbance is not a new phenomenon in Southern California with its long history of cattle ranching. Some biologists have argued that this has created an alternate ecological state. The degradation-of-mutualism hypothesis suggests that naturalized species appear to be poorer hosts for mycorrhizal fungi as well as less dependent on the fungi. Lower dependence and poor host qualities together create new habitats that are mycorrhizae poor. The theory predicts that this will create new stable states. The vegetation, with its prevalence of nonmycorrhizal fungal families, is now in a new equilibrium. Most important, these results also suggest that merely pulling out invasive plants and replacing them with natives is not enough. One needs to pay attention to the individual species and their relationships with the soil communities to fully understand the environmental context. Soil communities play an important role here and cannot be ignored.

Engaging Feminist Science Studies:
Notes from the Field

How should we understand the experiments and their results within the context of the social and feminist studies of science? In planning the experiments, I began to see the layered history of biology and politics embedded in the very structures of the experiment. And as the experiment progressed, my mind was churning with the complexities of studying natural systems. Of course, plants and animals are not people, and it is dangerous to read too much into the patterns of data. But as we have seen earlier, science is not innocent and removed from society. Deeply steeped in language and history, it is shaped by its historical contexts. These are worth examining. During the course of the experiments, several issues emerged.

TERMINOLOGY

The first issue we encountered immediately was terminology (Chew and Hamilton 2011). The loaded vocabulary of *native* and *alien* was troubling for all the reasons we saw in the previous chapter. Yet the literature and the theories of biological invasions were entirely predicated on these categories. We settled on *native* and *naturalized/introduced* as less sensational terminology and one that was historically in tune with the migration process. It was much easier to talk about naturalized plant species (since the term is often used in the literature) than naturalized soil (since soil is not one organism or entity), however. As a result, we often used the term *exotic* soil to refer to soil that naturalized species grew in.

LANGUAGE

The language of invasion biology creates a tautology—from beginning to end in experiments, the narrative is always about natives and aliens. It is impossible to escape this binary framing. Whether native and naturalized or something obtuse as Species Set A and Species Set B, the underlying formulation of experiments like ours is one of categories based on the purported geographic origin of the species. Any results that emerged would conceptually still be about that sole category of analysis. The very framing shaped the experiments and their interpretations. Any complexities that emerged were quickly and easy framed in the binary formulation of nature in place/nature out of place. This binary thinking forever locks in thinking about "the patterns of native/aliens"; it never moves beyond this formulation. It never allowed us, for example, to see if characteristics such as dispersal mechanisms, or reproductive strategies, or habitat heterogeneity, or other life history traits may be equally or more important. While this project did look at AM fungi, I was struck by how it was always folded into an analysis of AM fungi of native versus naturalized. The framing of native/alien structured all thinking and limited other possibilities.

There is also little doubt that this framing has shaped academic, popular, and activist frames. As we prepared the manuscript for publication, and if we were serious about its publication, we had to speak to the field—and this ultimately framed the story we needed to tell. And indeed the importance of framing is well documented in the literature on the sociology of academic fields. This work on invasion biology was funded because it built on a long history of research that has been constructed around the binary of native/alien. The work is read and synthesized by a community locked into the trap of this duality. Invasion biology has been a discipline of its own now since about the late 1980s (M. A. Davis 2009, M. A. Davis et al. 2011a). National organizations such as NSF, Nature Conservancy, as well as professional societies and journals recognize this as an urgent and important field. There are books, journal issues, and journals like *Biological Invasions* entirely devoted to this field. Money to fund these projects has been pouring in. For this work to be intelligible, and legible within the frames of the discipline, the binary framing is constitutive of the field.

More central to issues of native/exotic plants are questions of what gets to be called a "native" species. Given that the majority of Americans are immigrants themselves, the reinvention of "natives" as the white settlers and not "Native Americans" is striking. The systematic marginalization and disenfranchisement of "Native Americans" makes the irony all the more poignant. The love of the sequoia tree in the conservation movement is a eugenics symbol—its long life a testament to the glory of the white American settler as native. It is striking how plant and animal species never outgrow their labeling while the humans telling the stories (i.e., white Americans) have been rendered native in less than a few generations. But the categories of native/exotic are not as easy or clear-cut biologically speaking.[3] There can be dispute on how certain species can or should be characterized. While some use the term *invasive* more loosely to suggest the proliferation of a particular species, the official definition necessitates their alien origins.

Invasive Plant: A plant that is both non-native and able to establish on many sites, grow quickly, and spread to the point of disrupting plant communities or ecosystems. Note: From the Presidential Executive Order 13112 (February 1999): "An invasive species is defined as a species that is 1) non-native (or alien) to the ecosystem under consideration and 2) whose introduction causes or is likely to cause economic or environmental harm or harm to human health." (U.S. Department of Agriculture 2012)

In contrast to alien invasives, the term *opportunistic native plants* is used in some official circles to mark native plants that are "able to take advantage of disturbance to the soil or existing vegetation to spread quickly and out-compete the other plants on the disturbed site" (U.S. Department of Agriculture 2012). While natives are opportunistic, a character flaw, aliens are invasives, a serious threat! There have been many attempts to redefine the term, but it has become increasingly apparent that a rigorous definition of native and alien inevitably leads to contextual definitions, changing scales, and blurred boundaries (Warren 2007, M. A. Davis 2009, M. A. Davis et al. 2011a, Helmreich 2009). Throughout our project the ways in which the language framed and sometimes circumscribed the issue in dualistic frames of native/foreign was unavoidable.

ECOLOGICAL CONTEXTS

Invasibility, it emerges, isn't a characteristic of species; it has to be understood as a response to particular ecological habitats. At superficial glance the results reinforced all the ills of foreign species. Naturalized species degraded mutualistic fungi and created a context where native plants were at a disadvantage. They thus paved the way for a shift toward plant communities with nonmycorrhizal-dependent foreign species. But a deeper look at the data gave one pause. Is the foreign origin of plants the best ecological trait to focus on, rather than particular life history characteristics? In the floral survey, the one site that was undisturbed showed no significant differences in native and naturalized plant families that were nonmycorrhizal. Other studies have also found the critical importance of disturbance (Hobbs and Richardson 2011, Marvier, Kareiva, and Neubert 2004). Indeed, species that are "invasive" outside their native ranges are unlikely to be so within their home ranges (Hierro et al. 2006). Invasibility is thus a contextual response of certain species in certain environments. Disturbance and the history of grazing in Southern California seem key to the shift in the ecology of plant communities. Given the heterogeneity that native and naturalized plant species displayed, what if we focused on those characteristics rather than the native/foreign origins of plants? For example, biologist Mark

Davis has a useful suggestion in this regard (M. A. Davis and Thompson 2000, M. A. Davis 2009). Rather than classify species as native and exotic, categories that are often too heterogeneous and fraught with numerous complications, he suggests a more complex scheme attending to the varied life histories of plants and animals. He suggests three criteria: dispersal distance, uniqueness to the region, and impact on the new environment. Plants may be able to disperse their seeds short or long distances; plants may be common or novel in their uniqueness to the region, and their impact on the environment may be small or great. With this ($2 \times 2 \times 2 = 8$ categories) model, he argues that only two of them, those that are unique to the region and have short and long dispersal with great impact on the environment have the potential to become "invaders," as the term implies. We can move from a model where all exotic and naturalized species are seen as a problem to a model where we examine the ecology and life history traits of plants to evaluate their impact on the environment. Such a model that focuses on the biology of plants rather than their origins is a useful way to understand plant communities and their ecologies. In our experiments here, the ability of plants to be good hosts to mycorrhizae seems important to the ecological future of the site. Others such as Woods and Moriarty suggest we move from thinking of native and exotic purely as geographic categories to thinking of them as "cluster concepts," using multiple criteria, not just geography, to determine their status (Woods and Moriarty 2001). Indeed, we could see "invasion" as a symptom of larger ecological problems rather than as the ultimate problem that needs to fixed (Hobbs and Richardson 2011, Macdougall and Turkington 2005, D. M. Richardson et al. 2007).

Nature, biology, and ecology are never static, stable, and unchanging. Some have suggested that we move from a geographic label such as "native/foreign" to thinking in terms of a "damage" criterion—eradicate only those species that cause damage. Again, on deeper reflection, this is not so easy because sometimes exotic species do have both positive and negative effects. For example, the eucalyptus tree was introduced into the state of California from Australia about 125 years ago. In 1979 California decreed removing all exotic plant species capable of naturalizing and this included eucalyptus trees. While eucalyptus trees have plenty of negatives, it turned out that native monarch butterflies had grown dependent on the trees during their annual migration (Woods and Moriarty 2001). Indeed, it turns out that eucalyptus trees are now used by a range of butterfly and bird species, and at least one species of salamander has adapted to it (Woods and Moriarty 2001). In a world with profound human-induced movements of flora and fauna, a decision to suddenly decree what is native/foreign seems rather arbitrary. Local ecologies always evolve. Some naturalized species act as "native" and indeed some biologists see this as a criterion of naturalization (Carthey and Banks 2012). Returning to some notion of a yesteryear seems unproductive. Rather, we must wrestle with these complicated and at times confusing histories and be clear what ecological outcomes we are working toward—not a restoration to some nostalgia vision of the past.

Moving away from blanket labels and focusing on the dynamism of the biology of plants also allows us to recognize the complex nature-cultural pathways that have created the world as we know it. These categories are a lot more porous than most realize. Many species, for example, have become "culturally native" because of their long entanglements with countries and cultures (Warren 2007). Take, for example, a recent compilation of the one hundred heritage trees of Scotland. Of these, forty-two of them are technically alien species but have been rendered native through time. Often scientific judgments clash with cultural belongings (Rodger, Stokes, and Ogilvie 2003). Similarly, nearly all U.S. crops are foreign in origin, sometimes brought into the country through government-sponsored trips across the world. The current composition of plant communities of Southern California through cattle ranching is many hundreds of years old. Focusing on the biology of species, our recent ecological histories (including human ones) recognize the complex ways in which human beings have mediated the ecological shits. Shifting the focus from the invasive species as the problem to the complex geopolitical histories that have created our naturecultural world is a productive shift. Thus the solutions also become less about merely eradicating foreign plants and more about attending to the proximate cause of ecological shifts—overdevelopment, grazing, degradation of land, ecological fragmentation, soil erosion, and so on. If our desired model of nature, for whatever reason (biological, political, economic, or aesthetic), necessitates the management of particular species, we can do so without generalizing the pattern to all foreign species. A naturecultural approach also forces us to think more holistically about management. M. A. Davis's suggestion is one potential model among many that we can envision. While I am not necessarily advocating this model, attending to the biology of species allows us to refuse several binaries. The exotic/native binary seems more simplistic than the biology of species suggests. But we also avoid the nature/culture binary. Rather than analyzing this case purely as one of "cultural" production and therefore all hype, we can attend to the realities of changing ecosystems that are all too familiar and in some cases a cause for concern. If we choose to "manage" nature, we should realize, as the ecologist L. B. Slobodkin says, "that we preserve species and control species that are 'bad' from a human standpoint but with the understanding that the ethical problems are ours and not those of the organisms" (Slobodkin 2001: 8).

Toward Naturecultural Ecologies:
Naturecultures as Dynamic

In reflecting on the experiments, it is clear to me that biology can never be separated from society or nature from culture. The very language of native and alien, explanatory frameworks of invasive species, and the field of invasion biology with its self-propelling biological and political apparatuses are not easy to derail. Should Southern California go back to its pre-ranch days? Can it? Is this even possible? Or desirable? As such, naturecultures are not static but historically contingent themselves. The contexts of plants five hundred years ago, fifty years ago, and today are vastly different, leading to selections of different life history strategies. What if Southern California is in a new ecological equilibrium, as some ecologists believe? On what basis do we make decisions on what is "natural"? Paying attention to naturecultures allows us to recognize the intricate interconnections between natures and cultures in our analyses as well as our responses. These insights should force us to reconceptualize the biological and cultural language and frameworks by which we talk about the migrations and the redistributions of plant and animal species in our world today.

BEYOND BLAMING THE VICTIM

The language of invasive species misidentifies the problem that faces us and misplaces and displaces the locus of the problem. It scapegoats the foreign for a problem they did not create and whose removal will not solve the problem. The problem is not the foreign species per se but rather the human-made ecological disturbances that have caused ecological change to the plant and soil communities. Perhaps as some have argued, without naturalized and exotic plants we would be surrounded by impoverished, barren, and lifeless landscapes. We need to approach the problem with thought and reflexivity. The parallels between our approach to invasive plants and human terrorism are deep and worth pondering. With the rise of ideas and institutions of security and border security, we have also seen the emergence of the field of *bio*security, neatly tying into ideas of a nation in peril from its foreign humans and other biota (Hulme 2011, D. M. Richardson 2011). For those who believe that we should not steep ourselves in the return to false nostalgia of a white native country, why do we do so when it comes to our flora and fauna? If we are wary of pathologizing foreign humans, why not be careful about pathologizing foreign plants and animals? Instead, why can we not recognize the inherent diversity and dynamism of natures and cultures and work toward a vision that is ethical and respectful of all life?

Rhetoric of "natives" supports antidemocratic politics and ultimately yields less than maximally reliable sciences. "Naturecultures" force us to simultaneously attend to and transform both societies and the sciences that are dedicated to such projects, yielding more maximally objective and democratic results. Paying attention to the heterogeneity of native and foreign plants is also ecologically productive. The categories of native and foreign are not useful. Instead, let us focus on ecological traits that cause change we find undesirable and on the causes of these changes that we should be actually worried about—destruction of habitats, erosion of diversity, soil erosion, overdevelopment, monocultures, high input agriculture, pollution of air and water. And above all let us be clear that our choices are "human made" whether for economics, aesthetics, or enhancement of particular ecological characteristics like biodiversity, harmony, and species richness that humans have deemed important. Let us not fall back on age-old tropes of a pure nature or the natural.

We have to realize that nature is not that imagined nostalgia for a mythical yesteryear but rather an evolving entity, in which we are intimately involved. Whether we like it or not we are defining nature through our actions. This is not to fall back on an anarchic world where anything goes in the name of a free market or globalization. Rather, it is about taking responsibility for the world we live in and for us as a community to define the values that will guide us in our relationship with the natural world. We do not need to resort to the naive and powerful tropes of a fear of the foreign and alien or the calls for a nostalgic mythical white America. Instead, we should deal with the central vexing questions of variation, diversity, and difference that have plagued us for centuries in a larger context. Do we pathologize invasive species or recognize the valuable role they play in greening otherwise barren and desolate landscapes? Do we vilify foreign species or recognize the fundamental roles humans have played in redrawing landscapes from grazing and agriculture to more recent effects of climate change (Klinkenbord 2013)? This is the naturecultural world that can await us. If not, the ever-dynamic naturecultural world fueled with false nostalgia, irresponsible ecological management, overexploited landscapes, overdeveloped lands, and rampant consumerism will surely do it for us. The dire crisis of climate change, with its fast-changing plant and soil communities among many others, is surely all the evidence we need.

CHAPTER SIX

Aliens of the World Unite!

A Meditation on Belonging in a Multispecies World

Illegal aliens have always been a problem in the United States. Ask any Indian.

—Robert Orben

Every seed has a story . . . encrypted in a narrative line that stretches back for thousands of years. And if you trace that story, traveling with that little seed backward in time, you might find yourself tucked into an immigrant's hatband or sewn into the hem of a young wife's dress as she smuggles you from the old country into the New World. Or you might be clinging in the belly wool of a yak as you travel across the steppes of Mongolia. Or perhaps you are eaten by an albatross and pooped out on some rocky cropping, where you and your offspring will put down roots to colonize that foreign shore. Seeds tell the story of migrations and drifts, so if you learn to read them, they are very much like books.

—Ruth Ozeki, *All Over Creation*

The project on biological invasions that I have described in this section of the book was conceived as a joint project across the sciences and science studies, bringing the vast resources of the humanities and the biological sciences together to understand the natural world. After all, geographic origin and variation, alien and foreignness, are characteristics that can be shared by all organisms and objects. And yet, this project has evolved into something much more profound, an uncanny confluence of the academic and the social, of identity, experience, biology, culture, and politics. Geographies, genealogies, and biographies of variation have all been caught in a long historical debate about the politics of place. All the different parts of me, my work and my life, seem to be singing to each other in some grand polysymphony—multivocal, multidisciplinary, multidimensional, multispecies, multicellular, multiorganelle, multigenerational. This then is a meditation not so much about the unity of the world—of others and mine—but rather about the ways in which the lives and fates of so much of the world, natures and cultures, are entangled with each other. In this short reflection piece, I want to explore how academic, intellectual, political, personal, natural, and cultural voices came into conversation with each other. As scientists, if we reflect on our personal lives, it is usually about triumph or failure, or about passion, satisfaction, insight, determination, or perseverance. We hear about striving against the odds, the serendipitous discoveries, the rigors, pains, and joys of a life in science and the sometimes varied and inspiring biographies. Rarely, however, do we explore the profound ways in which these same narratives of personal, cultural, political, and economic contexts and the various life experiences or the trials and tribulations of one's life can have a profound and uncanny influence on scientific epistemology and methodology. We intuitively know it to be true and saw this vividly in part I of this book where eugenic concerns of the scientists profoundly shaped their scientific views. The cultural contexts of science are not just a byproduct or curiosity but in fact can be the source of new insights, new theories, and new knowledge for science. Studying science in context affords a glimpse into the entanglements of the natural and cultural, the personal and professional, and the political and intellectual, entanglements that become a rich site for new knowledge and theory making.

I came to this project with brave visions of interdisciplinary research. Alas, these always need to be conducted in a disciplinary world. I brought the tools from the biological sciences to examine the biology of invasive species and the tools from the cultural studies of science to examine the cultural contexts of invasion biology. I became the embodied site of interdisciplinarity. As the project progressed, the biologist and women's studies scholar in me watched, observed, and analyzed the evolving collaboration. As a brand new entrant into women's studies, I had just recently been introduced to the ideas of women, gender, and race and their troubled histories with science. And then a strange thing happened. An aspiring interdisciplinary scholar, I began to be struck by the growing synergies of the narratives of the alien and foreign plants I studied as a biologist and the emerging analysis of my own life grounded in ethnic, queer, postcolonial, and women's studies. Here then is a meditation on the affinities and affiliations of multispecies migrations in this contemporary moment of globalization.

I can still remember myself as the young girl who traversed the oceans to graduate school in the United States, straight out of college from India. Wide-eyed, exploring new geographies in a new country, a new world, new ecologies, new cultures, I could scarcely believe that someone would pay me to go to graduate school in the United States. Little did I reflect on the complex politics of the discourses of liberalism, third world development, and women's empowerment that helped me secure this opportunity. Never having left the country, I first had to secure a passport, that small booklet that declares one's national belonging, that marks the moment of legal and geographic belonging. The document also records one's exit from the geographic boundaries of the home nation and the entering of a new one. Armed with a passport, I visited the U.S. consulate to get official permission and a legal document securing entry, a visa to enter a new land. The usual questioning ensued. Why do you want to go? What is your purpose? Will you return? Yes? Are you sure? Student visa granted. A giant aircraft transported me across the oceans to a new land. Then on to the long immigration line in the airport for noncitizens of the United States. The passport was scanned through various security gadgets, piercing eyes stared intently at me as though boring into the depths of my soul. Passport accepted, visa acknowledged, and a stamp secured for a legal entry into the country. Then I had to confront the lines through U.S. Customs. Customs inquired, No living creatures? No fruits? No vegetables? No perishable foods? No visits to farms? Forms stamped and they let me go into the warmth of an American evening. Thus marked this legal process of migration of a human from one nation to another—from third world to first world. A sanctioned migration.

With a student visa, I completed my doctoral work. A postdoctoral opportunity arose. Then a job. So from a temporary student visa, to an H-1B visa, to a green card. From a legal alien to a resident alien to a permanent resident. Resident in one country, citizen of another. Then to a naturalized citizen, trading one passport for another, never ever to be a "native" again. Reflecting on these legal categories, I could not help but be struck by an analogous process among plants and animals. First the alien, the exotic, the foreign species, then the long-term resident, the exotic, and the naturalized species. In this age of global pandemics in many parts of the world, the border is a site of intense surveillance. If plant, animal, or human is suspected of an infection, some form of quarantine or denial of entry is often instituted. The desire for the "pure" migrant, vetted and sanctioned, is shared across species lines.

The parallels between my own life and my increasingly kindred plants and animals grew. I saw that the sustained campaigns to raise awareness about the danger and destructive nature of alien plants and animals were eerily similar to the xenophobic narratives against alien human immigrations. Decrying the constant influx of exotic and foreign species into the nation, these campaigns against foreign flora and fauna all highlight the erosion of native and local habitats, their economic and social cultures, and the destruction of nature. Similarly, anti-immigration activists campaigned against foreign humans, highlighting the erosion of native communities, their economies and societies, and the erasure, dilution, and destruction of native culture.

As we saw in chapter 5, there are many parallels in the ways in which foreign plants and animals and humans are treated. As the project unfolded my sympathies and kinship with these species grew. Books and articles have proliferated into a veritable industry against invasive and exotic plants, and the xenophobic rants against human immigration have repeatedly reached fever pitch. Living as a brown-skinned individual in the United States after September 11, 2001, the surveillance and suspicion is palpable. Are they so different from the "wanted dead or alive" posters for alien plants or insects? Are they so different from the penetrating eyes of the people on the road? Or the xenophobic rants or casual comments that pass you on the street? All based on one's presumed nationality or religious affiliation by phenotype or assumption. I cannot help but feel a deep sense of kinship with the equally stigmatized bodies of plants and animals. Thus, humans, plants, and animals are vulnerable to the new modes of surveillance where citizens are encouraged to alert a government agency by calling a phone number or sending an image (through an iphone app) of the potential danger. This mode of citizen surveillance seems deeply corrosive in a world in a perpetual war on terror. How do we keep those terrorists, those invasive species, out? We can do it by being suspicious of all foreign and alien creatures. The Republication candidates for president of the United States in 2012 appeared to be competing for who had the proposal for the tallest, widest, and most lethal electric border fence—of course, the border with the brown national neighbor, not the white one. How surprisingly deep are these multispecies affinities, these cross-species geopolitical kinship, and how uncanny and unexpected these networks of belonging emerging out of an interdisciplinary collaboration across biology and science studies!

And yet on greater exploration, I realized the many easy elisions that enabled me to feel like a kindred spirit to the foreign plants and animals. The idea of nativeness is developed from national states with historically demarcated geographic borders and boundaries with which they enforce who belongs and who does not. While it is true the native/alien definition in plants and animals comes from a nationalist spirit and a need to define national flora and fauna, there is, I discovered more than one way of defining the native. There are two forms of civil discourse—*jui solis* (right of soil or birthplace) and *jui sanguinis* (right of blood or inheritance) (Alonso 1995). Who is a native and who is a citizen and what rights each has depends a great deal on the national context and definitions of citizenship. Not all countries accord citizenship by birth within the nation-state. Even what is "native" requires a deeply contextual understanding. There are thus lots of different ways we can think about the native—through the geographic location of the origin of the species or taxon, through longstanding occurrence in a place, or through co-evolutionary or ecological relationships between the other species in the area. While each of these definitions shifts what we might consider native and alien, none get us out of the problematic notions of native and alien (Alonso 1995, Somerville 2005). It is through complex and fortuitous political histories that the plants, animals, and I, all find ourselves rendered alien in the United States.

Epistemologies of Kindred Subjectivities

What does it mean to increasingly identify as an alien studying aliens? How might such subjectivities, such alliances, be a source of productive knowledge production? There were several ways in which this shared subjectivity with alien plants and animals proved revelatory, enabling a deeper and a more expansive analysis.

The most profound insight was the recognition of the deeply entangled worlds of natures and cultures, that is, naturecultures. And indeed, the problem has been conceptualized as a fundamental question about similarity and difference. Do we want to live in a world where diversity and variation are celebrated or one that is homogeneous, monochromatic, and monocultural? Some have argued that we live in the epoch that can be called the "Homogocene" (Baskin 2002: 7). Some say that we should control alien species in the name of preserving diversity and variation. Just as cultural critics bemoan the McDonaldization of the world, so should biological critics bemoan biohomogenization by the entry of foreign species into native lands. Being against native species is not xenophobic but analogous to cultural critics asking to preserve rare languages or cultures. Being anti-foreign, they argue, is not the same as being xenophobic, but about preserving the "diversity of ecological assemblages" from the homogenizing forces of globalization (Hettinger 2011). As someone who has followed both the biological and the cultural critics, I am unpersuaded by both. This is a false binary and for the most part a moot point. Globalization has been ongoing for millennia, plants and animals have been moved around for just as long—hybridity and multiplicity is all around us. Also, globalization does not automatically dilute distinctiveness in either the cultural or biological world. New configurations, new identities, new possibilities emerge (W. O'Brien 2006). This is not to suggest that cosmopolitanism—of the floral, faunal, or human variety—is easy or unproblematic (Jamieson 1995, Pollan 1994, Soulé 1990, Paretti 1998). Blind cosmopolitanism, like blind multiculturalism, is likely to bring along its own sets of problems.

But this is not to suggest that globalization has been always good or productive. Movements of humans, plants, and animals have always been in reaction to particular political, economic, and social forces, and one can trace these circulations. Alien plants and animals share a common set of natural, cultural, and political contexts, and all of us are sometimes caught up in geopolitical circuits of power. As Karen Cardozo and I have argued elsewhere, some plants, animals, and humans share historical, economic, and political histories. Using the term *Asian American* as a geopolitical frame, we demonstrate how the term should be understood as one that is not just human but rather one that is multispecies (Cardozo and Subramaniam 2013). Like Asian American humans, Asian American plants and animals also share in its complex geopolitics, its colonial legacies, the eras of trade, and its cultural and culinary circulations. For example, the much maligned and invasive Asian longhorned beetle (ALB) owes its origins to China's policies to combat soil erosion and deforestation, which resulted in the country planting massive rows of monoculture poplar plants as wind-breakers. The ALB favors poplar trees, and these monocultures allowed an explosion in growth of ALB populations. Soon after, when trade exploded in the 1970s and 1980s and there was need for wooden crates, China was well equipped to provide poplar wooden crates, which of course now carried the larvae of ALBs. These wooden crates were shipped worldwide, enabling the extensive circulation of these beetles. We need to remember the current problems with ALB are part of these complex circuits of ecological trade, environmental, and commerce policies (Alsop 2009, MacAusland and Costello 2004). And this is the context we should recognize as we are inundated with "wanted" posters in newspapers, magazines, movie theaters, billboards, and i-phone apps that call for its eradication and destruction.

The much famous Georgia peach also has its origins in China, being brought to the United States through the travels of USDA agents (Kaplan 1991). Of course, we want to domesticate and appropriate the peach as our own, even calling it the Georgia peach, while the ALB is remembered for its destructive foreignness. We can tell similar stories of many other plants and animals, as we can about humans. Thus all of us, creatures of this earth, are caught up in complex and unexpected networks of entanglements.[1] As chapter 5 shows, it is astonishing sometimes how the rhetoric, ideology, and politics from one sphere quickly engulf another. The xenophobic rants of anti-immigrant activists can quickly be heard in environmentalist circles. It is a keen awareness of these kindred subjectivities, of broad interdisciplinary approaches, that allows us to trace these dense and vibrant circulations.

Nationalism, Race, and Xenophobia: Disciplinarity and the Circuits of Knowledge

Just as the geopolitics of Asia and America have generated a richly textured world of multiple species of Asian Americans, so have other forms of geopolitics. Conversely, this work allows us to recognize the limits of disciplinarity. The narrow lenses of the biological sciences see invasive species purely as a "biological" problem, and the narrow lenses of the humanities see xenophobia or racism as a human problem. Each fails to see how their disciplinary frames need be so much broader and more complex. There are deep links between the nationalization of nature and the naturalization of nation (Sivaramakrishnan 2011). In discussing the case of Europe, Zimmer notes that "as politicized nature, particular landscapes evolved into integral parts of historicism's search for national pedigrees, something that happened across Europe in the late 18th and early 19th century" (Zimmer 1998: 641). Tracing the rhetoric of invasive species brings the natural world squarely within race and immigration politics. As we have seen before, it is not accidental that historical moments of strong xenophobia in human cultures have been associated with panics about foreign plants, animals, or germs. Indeed, Olwig argues that such ideas of nationalist landscapes/cultures are drawn from a common epistemological template that can be traced back to the Renaissance, when methods of surveying and cartography were rediscovered (Olwig 2003). The links between plant/animal control and human control are well documented; perfection in gardens and peoples is rooted in an ongoing struggle against "difference" (Mottier 2008). And the vast resources of the humanities have amply demonstrated how Hitler's vision of pure human populations was accompanied by visions of pure gardens,[2] demonstrated in this quote by the German landscape architect Willey Lange:

Our feelings for our homeland should be rooted in the character of domestic landscapes; therefore it is German nature that must provide all ideas for design of gardens. They can be heightened by artistic means, but we must not give up the German physiognomy. Thus, our gardens become German if the ideas for the design are German, especially if they are borrowed for the landscape in which the garden is situated. (trans. and qtd. in Groning and Wolschke-Bulmahn 2003: 79)

As Rodman notes, applying the native/alien binary runs the risk of "the precarious utopia of a racially pure Reich" (Rodman 1993: 152). In a context of racism, xenophobia, and nationalism, the ideology of "blood and soil" made deep links between pure humans and pure gardens. The ideas of national cultures and gardens emerge again and again and have appeal at various times across countries (Groning and Wolschke-Bulmahn 2003).[3] And indeed, these kindred analogies between humans and plants were also embraced by American landscape architects such as Jens Jensen, Wilhelm Miller, and Frank Waugh.[4] For example, Jensen is quoted in a 1937 article as saying:

The gardens that I created myself shall . . . be in harmony with their landscape environment and the racial characteristics of its inhabitants. They shall express the spirit of America and therefore shall be free of foreign character as far as possible . . . The Latin and the Oriental crept and creeps more and more over our land, coming from the South, which is settled by Latin people, and also from centers of mixed masses of immigrants. The Germanic character of our race, of our cities and settlements was overgrown by foreign character. Latin spirit has spoiled our lot and still spoils things every day. (qtd. in Groning and Wolschke-Bulmahn 2003: 85)

The fact that American eugenicists pointed explicitly to the forest and gardens of Germany as the birthplace of the instinct for democracy is a chilling reminder of the origins of U.S. nationalism and nativism. The idea of "native" nations and plants and animals has a long history and is deeply embedded in a politics of purity. It should come as no surprise that with the rampant fear of immigration in the United States, there was a 53 percent growth in housing units in gated communities between 2001 and 2009. "Stand your ground" laws show one response to the calls for assimilation (Benjamin 2012).

Thinking natureculturally on the politics of purity, I see the connections of xenophobic rhetoric of invasive species to those that sometimes permeate discourses against new reproductive technologies, genetically modified organisms, or the movements for local foods. One of the reasons cited against foreign plants is preservation of the "genetic integrity" of native and nations. The fear that foreign species might interbreed with natives yields discourses that have a familiar ring of anti-miscegenation policies (Smout 2003). Yet, as Forrest and Fletcher argue, when pressed for a definition of genetic integrity, "there is generally some reluctance on the part of those employing such terminology to come to the point, although emotive issues connected with the archival 'preservation of our priceless heritage' and perhaps a variety of 'ethnic cleansing' are seldom far from the surface" (Forest and Fletcher 1995: 99).

ROOTS OF COINCIDENCE: THE POLITICS OF PURITY

Fears of impurity usually grow alongside fears of pernicious sexuality, global miscegenation, and unbounded migrations. And indeed, living in the United States, I began to note the politics of fears in multiple sites, all feeding on familiar tropes of race, and nation (B. Hartmann et al. 2005). What is fascinating is that despite deep ideological differences, in three current issues— invasive species, GMOs (genetically modified organisms), and NRTs (new reproductive technologies)—positions of the political left and the right converge. Whether some individuals are looking for a return to an imagined nostalgic past or for a future without foreigners, individuals across the political spectrum make similar arguments. Some environmentalists want a pure nature. Some feminists are critical of reproductive technologies and their impact on women. Some religious conservatives are afraid that we are taking the place of "God." And some conservative environmentalists are afraid that we are destroying God's creations. Conversely, some environmentalists are critical about the purity discourse while conservatives embrace technology and the free flow of flora and fauna in the name of free markets and the free flow across borders. What do we make of such a convergence? Are the roots of these fears and anxieties the same? Does the common rhetoric belie an anachronistic political similarity between the right and the left? Or is it entirely coincidental? Briefly, I want to suggest four ways in which I believe the arguments of the right and the left converge as they express their opposition to alien biota, GMOs, and new reproductive technologies.

First, the creation of the "other." In each of the three cases, the resulting product—the proliferating invasive species, the genetically modified organism, or the technologicalized mother/baby—is viewed with deep suspicion and as "foreign." In the case of invasive species, the term *invasive* literally becomes synonymous with exotic/alien/foreign species—ruling out the possibility of native invasive species, which seldom get any publicity. Signaling the ultimate monster, GMOs are often even dubbed "Frankenfood" (Egan 2011). The many "accidents" of new reproductive technologies—where white women carry black babies, black women carry white babies, grandmothers are pregnant with grandchildren, and women routinely carry multiple pregnancies—all warn of the "bizarre" and the creation of possibilities that are unlikely to occur in our peculiarly gendered and raced world. It valorizes women's bodies as the sacred site of motherhood. In each of these cases, the language "reinscribes" particular notions of the "other," simultaneously reinforcing the normative as the native, the natural, the pure.

Second, the "other," most often the female, is often attributed with "hypersexual" fertility. Invasive species are routinely ascribed with superfertility. Consider, for example, the title of an article on invasive species: "They Came, They Bred, They Conquered." Within the GMO literature, one sees the fear of transgenes quickly moving to other plants and even crossing species boundaries to create "superweeds" with superfertility. The case of NRT is more complex and interesting because the object of superfertility and primary beneficiary of NRT is the white woman—whose fertility our culture deeply desires.

Third, linked to this is a valorization of nature and the "natural." Fundamentally in this framework, "nature" is a realm that is seen as removed from human interference but also human-friendly, safe, and trustworthy, that is, products of nature are safe for humans. By tinkering with nature, humans are argued to assume an unparalleled arrogance and are accused of playing "God." In this vision, if respected and left undisturbed, nature nurtures native species, nature produces "wholesome natural products" with pristine seeds that are good for you. If, however, we disturb this co-evolved nature with its own checks and balances, we are at risk of unleashing monsters. Invasive species transcend the "natural" order by moving where they do not belong. GMOs enable unnatural gene mixing and NRTs threaten the "natural" process of women and reproduction. Interestingly, companies that produce transgenic plants and GM food have moved to use the same rhetoric to celebrate GM food because it will allow the "natural" to be more "natural." They suggest that producing varieties with higher yields will save biodiversity and ultimately conserve native forestland. Similarly, transgenes that bring pesticide and weed resistance to plants will reduce the use of pesticides and herbicides and ultimately create more "pure" nature and sustainable agricultural practices.

Fourth, the rhetoric of purity is striking in each of these discourses—"pure" nature, "pure species," "pure women," species fidelity. Anxieties abound about native and exotic species cross-breeding, thus "contaminating" the native gene pool and gene purity. The rhetoric emphasizes purity by highlighting "leaking genes," "genetic pollution," and "contamination." Activists and policy makers have created a purity index and have developed standards to measure "seed purity."

Similarly, the vast resources of biology also remind us of the complex interactions that make ecosystems. We have observed the profound impact of ecological managements when new species have been introduced to control pests or weeds and have caused more harm than good. Or at other times when foreign species have evolved to form new communities where native species have become dependent on the foreign species (Woods and Moriarty 2001). Disciplinary thinking fails to illuminate these interconnections and entanglements.

The same sentiments of purity inhabit disciplines as they systematically dismiss methods, methodologies, and theories from outside their disciplines as trivial or sloppy. Thinking natureculturally and interdisciplinarily unleashes the vast resources of the humanities, arts, social sciences, and sciences. It gives us access to the vast repertoire of tools, theories, methods, and histories. It opens up our imagination to the vast possibilities of the universe—of poets and naturalists, of fiction writers and science writers, of rhetoricians and physicists. It reminds us yet again that words and language are powerful—not transparent and apolitical but powerful tools that can be used toward divisive and violent ends or toward egalitarian, peaceful ones. For those working toward a better world, thinking natureculturally opens up the insights of conservationists and social activists to each other. It allows us to see that the multiple strategies of various social activists—pro-immigrationists, ecologists, humanists, internationalists, and globalists—can be learned and shared.

The Politics of Assimilation

Alongside a politics of purity, one also begins to see a call for a politics of assimilation. Biologists have long recognized the "naturalization" process of plants and animals and that many foreign species become "culturally native," deeply implicated in local cultural geographies. Some of the calls warning about the dangers of a rigid native/alien distinction come from those who recognize the positive contributions of many foreign species. In addition to many crops, vegetables, and economically important plants and animals, alien species have often naturalized into the biotic world in productive and intricate ways. The case of the eucalyptus tree that is now important to native monarch butterflies as well as several bird species reminds us that a vast number of foreign species have been in the United States for centuries, even millennia. These organisms have evolved to create new communities and new biotic interactions, at times creating new equilibria. The wholesale eradication of alien species is particularly unproductive in the face of climate change that is fundamentally transforming our Planet and local environmental contexts. What does it mean to harp on a return to a past when the environmental contexts the plants evolved into no longer exist? It would appear that we need to move away from a clear separation of culture and environment and toward thinking in terms of relational geographies of plant/human interactions as global environmental change refigures circuits of plant, animal, and human migration, adaptation, adaptability, ecology, and evolution (Head and Atchison 2009).

All of these factors have pushed many to contemplate a less rigid and more flexible approach to the environment, driven less by arbitrary categories of native/alien and more by the empirical realities on the ground. In recent years, a productive site of dealing with invasive species has to do with modifying and expanding our culinary habits. Environmentalists have been getting people to combat invasive species by eating them. For example, in recent years they have tried to harvest crayfish in Lake Tahoe in order to improve the water quality through commercial harvesting. In Nevada, mostly made up of desert, availability of local seafood is exciting. "This is where science stops and you need people to step in and make a decision to improve the lake," a local scientist argued (Onishi 2012). While Asian carp, a delicacy in China, has been overfished there, it is reviled in the Midwest. Companies in the Midwest are now exporting carp to China (Frazier 2010). Locavore movements are popularizing recipes of invasive species (often considered delicacies in their homelands) within the United States. After all, Chilean sea bass, now a delicacy, was less palatable as the Patagonia toothfish! In similar moves, companies in Illinois are trying to convert carp into organic fertilizer as well as introduce products where fish meat is ground into products such as salami, bologna, and even jerky (*PBS Newshour* 2012). A recent *New York Times* story featured Ms. Wong, who having given up fighting her weeds, has instead turned to eating most of them. Moving beyond the narrow offerings of commercial vegetables and herbs, she has taken to growing weeds to wide acclaim and now is popularizing them through a new cookbook (Raver 2012).[5]

A more spiritual take on invasive species comes from those who see invasive species as healing a destroyed Planet. Arguing that many of the species considered invasive are in fact medicinal plants in their home country, they argue that invasive plants are in fact a great resource and boon. Rather than respond with toxic pesticides to control them as some restoration ecologists do, invasive plants can grow on damaged land and perform an essential ecological function to heal both the land and the humans who live on it. In such a view invasive plants transcend the good/evil binary to become the healers of a damaged and sick Planet and world (T. Scott and Buhner 2010).

The Politics of Knowledge and
Knowledge Production

Is paying attention to this kinship between my foreign status and that of my objects of study dangerous? After all, isn't objectivity one of the key cornerstones of science? Despite all the claims of objectivity, we have seen again and again in history that science is far from objective. As we have seen in the introduction to this book, a vast literature in feminist science and technology studies shows us that objectivity is an illusion, a mirage that obscures the ways in which science is deeply embedded in its historical and political contexts. That said, a kindred subjectivity is not about a relativistic world where everything goes or where subjectivity comes to stand in for some idea of absolute truth. Rather, thinking natureculturally, thinking with and through kindred subjectivities, forces us to think reflexively. It enables interdisciplinary thinking, compelling us to consider the complex circulations of knowledge. It moves us away from dualities such as labeling all of one category of plants and animals as "evil." Rather, it allows us to understand how we have all come to this country because of complex histories. It allows us to understand that weekend campaigns to go pull out the latest undesirable species from the local pond will not solve the underlying problem. It forces us to acknowledge that we have to think more broadly about national and international environmental policy and how our local problems are connected to larger national policies of energy, development, and globalization. We need to work against an unproblematic scientism in environmental policy and take seriously the institutions of governance, systems of values, and ways of knowing (Jamieson 1995).

The world itself is not disciplinary—plants, animals, and humans are connected through complex histories and geographies. The interdisciplinary experiment I was involved in allowed me to show how moving past disciplinary thinking is helpful in our interdisciplinary interconnected world. I was able to transfer ideas across disciplines to explore the power of language and rhetoric; to bring insights from colonial and postcolonial studies into botany and zoology; to trace the historical, geographic histories of plants, animals, and humans simultaneously; to stretch narrative theory into telling new stories about our nonhuman co-inhabitants. Understanding this means opening ourselves up to a naturecultural world and the vast resources of interdisciplinarity. We cannot understand it any other way. Challenging the nationalist nativist landscapers we read earlier, the Jewish writer Rudolf Borchardt, who was persecuted by National Socialists and who wrote this in 1938, puts it particularly well:

If this kind of garden-owning barbarian became the rule, then neither a gil-lyflower nor a rosemary, neither a peach-tree nor a myrtle sampling nor a tea-rose would ever have crossed the Alps. Gardens connect people, time and latitudes. If these barbarians rule, the great historic process of acclimatization would never have begun and today we would horticulturally still subsist on acorns . . . The garden of humanity is a huge democracy. It is not the only democracy which such clumsy advocates threated to dehumanize. (trans. and qtd. in Groning and Wolschke-Bulmahn 2003: 86)

All this is not to suggest that a nonimmigrant could not come to these insights. Indeed, many have. Nor is it to suggest that all immigrants would see these connections. Many do not. Rather, it is to suggest that our life histories, our experiences, our identities, that the vagaries of life can at times open up the world in particular and surprising ways. We should regard these as wonderful and rich opportunities to understand the world in new ways. I would never have seen these connections, never have been able to explore these connections, if not for the rich opportunities of a broad and interdisciplinary training in biology, environmental studies, feminist studies, ethnic studies, critical race studies, postcolonial studies, and queer studies. If our training in the humanities and the sciences could train our eyes, ears, and minds to be open to the world outside the discipline, our theories and knowledge about the world would be that much richer. Scientists and humanists are not removed from the context of their lives. Denying these connections is not objectivity, but rather a lost opportunity.

Toward a Multispecies, Multidisciplinary View of Life

If we open ourselves to a naturecultural world, we have new stories to tell, new narratives, new histories, new cartographies of knowledge. Listening to poets and scientists alike, the naturecultural world is teeming with insight and possibility. The atoms and subatomic particles connect all life and nonlife in this universe. Life on earth is connected through complex evolutionary histories, literally sharing a material connection through the helices of our DNA. And indeed, we are not individuals but multispecies entities ourselves, each of our cells a collection of multiple species that have over the centuries symbiotically evolved to create individual cells and subsequently whole organisms (Margulis 1998, Marguis and Sagan 2002). Yet, our biologies and cultural ideas resolutely center the individual (Gilbert, Sapp, and Tauber 2012). And what about science? We tell a unitary story of "western science" as a linear, progressive story of knowledge produced exclusively in the west (Teresi 2001). Yet, science itself is the product of multiple miscegenations. Appropriating, embracing, and accumulating knowledge as it traveled through complex histories of colonialism, trade, and empire, science is the ultimate mutt. These pedigrees erased and forgotten, the mythologies of a pure "western" science continue to be told in colonial and colonized worlds. We can bemoan the colonial legacies, but if we continue to tell these stories of some pure entity called "western science" in the twenty-first century, we should also see it as an unfortunate and lost opportunity to tell new stories.

These are the mythologies we tell about science, about ourselves. Divided always, specialized always, each neatly packaged in our disciplinary boxes. Yet, we are connected through molecules, through the helices of our DNA, through histories, through geographies, through colonial travel, through vibrant trade, connected through an ever global, ever connected world. Oliver, a character in Ruth Ozeki's *A Tale for the Time Being,* creates an art project, a botanical intervention he calls the Neo-Eocene, a collaboration with time and place. He argues that the rapid onset of climate change will radically expand the term *native* to include formerly and even prehistorically native species. After all, native depends a great deal on how you define it. We might well have a return to older times!

Like the peach, the kudzu, the snakehead, and the carp, I find my own journeys and travels caught within these circuits of global capital. My own value and worth is tied to vast naturecultural assemblages. I began with noting how in moving from India to the United States, I was transformed from a native to an alien. Yet, my exotic status as the only South Asian graduate student in my department twenty years ago is now a rarity. Today, South Asians and Chinese are overrepresented in science and engineering departments. What was exotic once is now being transformed through the new geopolitical realities into the fear of India and China, the emerging powers of the future. My identity, my fortunes, and indeed all of ours are tied to these global networks and circuits of power. We cannot escape them. Aliens of the world unite! We are all aliens, we are all natives, in this vast entangled naturecultural multiverse.

PART III

Biographies of Variation
The Case of Women in the Sciences

The limits of variation are really much wider than anyone
would imagine from the sameness of women's coiffure
and the favorite love stories in prose and verse. Here and
there a cygnet is reared uneasily among the ducklings
in the brown pond, and never finds the living stream in
fellowship with its own vary-footed kind. Here and there
is born a Saint Theresa, foundress of nothing, whose
loving heartbeats and sobs after an unattained goodness
tremble off and are dispersed among hindrances instead
of centering in some long recognizable deed.

—George Eliot, prelude to *Middlemarch*

CHAPTER SEVEN

Through the Prism of Objectivity

Dispersions of Identity, Culture, Science

Although the universe is under no obligation to make
sense, students in pursuit of the Ph.D. are.

—Robert Kirshner, qtd. in Ferris, *The Whole Shebang*

| Haha!

There was an arid precision to life that bothered me. In
fact, science was like table manners, a ritual correctness
in life where morality yielded to routine.

—Shiv Vishwanathan, "The Laboratory and the World"

I guess transformations—the really important ones—
require more than time and distance, and even desire.

—Chitra Banerjee Divakaruni, *Arranged Marriages*

My memories of second grade largely come from a wastepaper basket. In my second-grade class, whenever a student talked too much, usually when the teacher's back was turned and there were more interesting things to talk about than the lesson at hand, the teacher would empty the wastepaper basket. Then she would make the student stand in the wastepaper basket in the corner of the room to observe the rest of the class. My year in second grade is filled with images from that corner, my feet in the basket. My teacher added a comment to my report card, "too talkative." Despite the fact that I was being disciplined, my self-confidence and self-worth flourished from the wastebasket corner as my friends would make faces and signs at me when the teacher was not looking.

This sense of creativity, self-confidence, assertiveness, and rebellion from boredom that punctuated my educational experience in India is one I pondered often during my years in graduate school. I ascribed it to a sense of privilege I felt as a middle-class Indian girl born into an upper-caste family belonging to the religious majority, schooled mostly in urban all-girls' schools. As a member of the majority, all social markers worked in my favor. Marginality—minority status in ethnicity, nationality, religion, caste, educational background, the "otherness" that breeds insecurity and marks one's body and identity—were alien to me. Entering a graduate program in biology in the United States, I literally became that alien, officially, a nonresident alien. The science classroom in my graduate years contributed to a growing sense of marginality, insecurity, and invisibility. I desperately wanted to conform and belong. I struggled to find a voice to articulate a word, a sentence. I longed for my wastepaper basket years!

When I began taking courses in women's studies as a graduate student in the sciences, the experience was completely different. The very social categories that marked my "otherness" in the sciences were now tools of knowledge making, providing rich standpoints from which to interrogate U.S. culture, academia, feminist theories, and the educational settings I found myself in. These new perspectives were enabled by theories and contexts of feminist scholarship and the pedagogical tools in these particular women's studies classrooms. Feminist scholarship gave me the tools by which to interrogate my experiences on the other side of campus. What is it about the U.S. graduate education system and the sciences in particular that can turn a relatively self-confident third world woman into an insecure, marginalized one? How are race, gender, nation, and sexuality coded into the enculturation of scientists and the educational process?

During my graduate training in the sciences, I was an avid journal writer. In this chapter I use these memories to explore and theorize what these experiences can tell us about the culture of science. To be sure, experiences are at once rich in their depth and detail yet particular and idiosyncratic as a source for theory. To be sure "experience" is by now a fraught and much critiqued site for theory making in feminist scholarship (Joan Scott 1991). In using my experiences in graduate school, my goals are less about codifying an authentic or definitive narrative of being a woman in the sciences. Experience is not the "origin of our explanation, but that which we want to explain. This kind of approach does not undercut politics by denying the existence of subjects; it instead interrogates the processes of their creation" (Joan Scott 1991: 797). In this book, I am interrogating the processes that shaped my experiences. I explore questions of genealogy, geography, and biography and their interrelations in the sciences through my experiences in the hallways of science. These experiences, while individual, point to how the histories, cultures, and epistemologies of science are operationalized within scientific life reproduced through the generations, sometimes with surprising fidelity. My decision to use this form comes from three main gaps I see in feminist science and technology studies. First, personal narratives and memoirs are a highly underutilized mode in feminist science studies. Personal narratives can reveal the personal, professional, and institutional connections that are underdeveloped in the field.[1] Women are a diverse group, and I explore differences among women of color with the complexities of gender, race, class, and nation. Second, feminist science studies has largely focused on the construction of scientific knowledge rather than the production of scientists (Traweek 1992), but I see value in doing these simultaneously. To me, feminist frameworks and critiques of objectivity were immensely useful as I negotiated my life in science. I explore how presumptions of objectivity and rationality constrain and curtail originality and innovation in the culture of science, shaping normative expectations and scientific norms. Feminist scholarship and its critical reflexivity can enrich the lives and lived experiences of scientists as well as science.

Finally, and most important, I want to argue that graduate education is a critical juncture in the educational ladder where students are "encultured" into their professional identity as scientists (Subramaiam and Wyer 1998). Graduate education is not only about learning scientific methods and methodologies, but also about developing a professional identity. After all, graduate school is structured around enabling a remarkable transformation in one's relationship to the faculty—from "student" when one enters to a "colleague" when one leaves. Yet the power dynamics that inform this transformation are hardly benign though seldom discussed. Graduate education is an undertheorized and relatively unexplored area, a rich site through which to understand how scientists are trained and produced.

It is striking that in higher education the term *selection* is used in reference to sorting students. In undergraduate education, large introductory science courses are often called "weeder" courses—where weak students are weeded out so as to "select" for the scientifically gifted students (Barton 2001). Similarly, in graduate school, ideas about who is "gifted" or can "cut it" or has the "spark" are rampant. Still, being academically gifted does not ensure the makings of a great scientist—something else is necessary—and so my graduate student colleagues and I often pondered the mythologies about this *je ne sais quoi* factor. It was striking to me how in my graduate program the frame of "natural selection" of plants and animals permeated similar frames of the "selection" of students. There is thus a deep structural resonance between our genealogies of variation (the research practices that produce certain science and scientists), the geographies of variation (the processes that brought new species into the

161

west and helped the global circulation of floral, fauna, people, and knowledges), and the biographies of knowledge (the practices that have "selected" a strikingly homogeneous scientific institution and ivory tower). As I hope is clear, genealogy, geography, and biography are part of a "naturecultural selection" that shapes the sciences—its practitioners, cultures, and knowledges. George Eliot in the opening epigraph to this section reminds us of the many brilliant women, people of color, and third world individuals who were excluded from science and scientific inquiry and of those who persisted, how their achievements were rendered invisible or marginalized in "lost history." Anyone who has worked with graduate students today will recognize the persistence of these patterns. The many ghosts of talented individuals roam the hallways of science. We need to listen to them carefully.

What follows represents my efforts to hear the ghosts, prompted by flashes of memories captured in my graduate school journals.

Beginnings

I have arrived! The excitement of leaving one's country is tremendous. The dream of enrolling in a graduate program in the sciences is unfolding before me. I am all eyes and ears. Large, sparkling buildings, well-equipped labs, and exciting research. It is amazing to be in the heart of the scientific enterprise. On the social front, everyone smiles and says "Hi!" Coming from a city, the friendliness is infectious. I feel loved and accepted. It is wonderful. Of course, they never get my name right. Oh why, oh why do I have such a cumbersome name! They all tell me in astonished tones how wonderfully I speak English. I am puzzled by what their impressions of India must be.

I happened on the United States with what all scientists dream of—a null hypothesis! I believed that there was no racism, sexism, and classism here. I came into a biology department where the number of men and women graduate students had been equal for more than a decade (although not at the faculty level). Thinking back, I had a naive, exuberant openness that I still find incredible yet heartening. My training in the sciences in postcolonial India was strongly grounded in assumptions of objectivity, rationality, and value neutrality. Thanks to a traditional education (British style), I fervently believed that the sciences provided a haven where identity ceased to matter. I reveled in the possibility of belonging to a profession where social and cultural prejudices were irrelevant and where hard work and merit would win the day. I was an individual like every other. Such was the legacy of colonialism in British India, the enlightenment postcolonial subject!

Growing up in post-independent, secular, urban India meant growing up with the promise of science. Science was central to my image of modernity. Science, as it was taught to me in school, as it was represented in the books I read and the popular culture I watched, was "western" science. Indigenous forms of science and medicine had not been (and are still not) integrated into the rubric and authority of science. Even though Indian scholars have made rich contributions to western science and mathematics, I was never taught this. Religious orthodoxy was in my eyes associated with discrimination, "backward" thinking, superstition, and blind faith. When my family would consult the astrological charts to look for auspicious times for a move or tell me I should not sleep with my head facing north, I scoffed at them. When I saw families separating girls and women during their menstrual days, I was outraged. I ridiculed silly superstition, laughed at irrational tradition, and became enraged when I saw discriminatory or hateful practices against any man or woman. Growing up in postcolonial India meant having access to a vibrant and visible feminist movement. To my young, modern, and urban feminist self, my feminism and politics were linked to ideas of modernity, and modernity was linked with claims of reason, and reason was linked with objectivity and rationality of science. Science could not have found a more committed or ardent citizen!

Very early in life, I was passionate about the sciences. I was drawn to their call for logic, reason, rationality, and objectivity. Science was a meritocratic world where my identity as a woman, Indian, third-worlder was irrelevant. The white men (dead and alive) who inhabited my textbooks were my role models, and I was quite oblivious to my brown skin or my sex. A large poster of Charles Darwin hung above my desk. It did not occur to me that, with the exception of C. V. Raman and J. C. Bose, there were no Indian men in my science textbooks and certainly no Indian women. The hope for the world rested squarely with science. It came as no surprise that after an undergraduate education in biology in India, I should cross the ocean and come to the United States for a graduate degree in evolutionary biology full of visions and dreams of being a model scientist.

Throughout postcolonial India, "western science" was the science that the Indian state supported; alternate forms of science and medicine have remained in the periphery. As Susantha Goonatilake suggests, modern science in the third world has always been defined by the center, that is, the west, and any creativity that has emerged has come from indigenous and peripheral practices (1984). Western science had been transplanted into India and embraced as the central force for modernity and development. For a young woman who grew up in such a climate, being part of a graduate program in the sciences in the west was a dream fulfilled. I had indeed arrived!

Enculturation

AN ODE TO THE FURNITURE WORLD

It has been a month. I cannot put my finger on what I am doing wrong, but I have a deep sense of loneliness. People tell me about culture shock. When people can talk to each other, what can go wrong? It must be me. No one around seems to understand the alienation I feel. It is this funny combination of feeling invisible and sometimes too visible! How should I act? Who should I be to fit in? I solve it by a tremendous affinity for the furniture world. So I have begun to play this game every day where I become the table, the chair, the wall, or any inanimate object that catches my fancy, and watch and absorb everything—gestures, words, phrases. It is all new. There are assumptions, expectations, which I cannot fathom or figure out. Being entirely stripped of all context is a very frightening experience. On the social front, all conversations are new information. So, I listen, absorb . . . We go out to eat. I have no clue what is on the menu and no one around me seems to have a sense of my predicament—the ability to eat spaghetti is not an inborn human trait! And so yet again, I watch, emulate, absorb . . . Then, there is living in a technological society. Coke machines, copy machines, bank machines—I stare stupefied, paralyzed. Then comes the absolutely awesome experience of walking into an American grocery store. Shiny fruit and vegetables—all sterile, waxed, and polished (which I am discovering is what they taste like). The choices are mind-boggling . . . I have become quite good at it. The immobility, the invisibility, are like second nature. I have a strong solidarity with all that is wood or steel!

At one level, these sentiments are likely shared by many foreign students. Encountering a new culture, a new educational system is decidedly and necessarily disorienting. But having talked to students across the disciplines, I'm convinced that the particularities of scientific culture differ from those of other disciplines. In hindsight the framework of science that I brought from middle-class urban postcolonial India was unlike that of my U.S. counterparts. Many of them had worked in labs, read scientific papers, and done research projects before they came to graduate school. The world of science was not entirely a figment of their imagination fed by Jacques Cousteau, Richard Attenborough, and science fiction movies! Furthermore, most had taken courses in the humanities and the social sciences as undergraduates. One of the legacies of British colonialism was an educational system that specializes very early. After tenth grade, I chose to follow the "science" route (as opposed to arts or commerce). During my undergraduate years, I chose a "zoology" degree. With little background in the humanities or social sciences, I had no theory of culture. Watching and emulating others would do the trick. These were superficial things—nothing to do with science. I assumed that my sense of loneliness and cluelessness in class had more to do with my educational background and capacity than with the culture that shaped the social relations I was part of—who talked to whom, who helped whom. There was little recognition that such social networks would be crucial to learning. As I discovered, graduate and higher education, at least in the sciences, has less to do with book learning than with intellectual community. Community networks can make or break you. In this, I was ill prepared for graduate school because I presumed my knowledge of nature was all that mattered. The graduate program I entered was ill prepared for me, as well. There was a kind of cultural heterogeneity—there certainly were a lot of different people getting along with one another. After all, the conservative Texan felt he had little in common with the Northeast liberal, but nevertheless they worked together. All in all, I came from a little farther away. While the graduate students were almost exclusively white and from the United States, the wide U.S. geographic differences reinforced the robust rhetoric that difference was irrelevant to the pursuits of science. Initially, I did not realize the degree to which assumptions about the third world, and visions of "the Orient" in particular, shaped their comments and their impressions of me. But at one point I had to recognize that I felt exotic/"other"/marginal in constantly being asked to be the spokesperson for another part of the world. In hindsight, the level of ignorance (theirs and mine) is rather striking. There was no vocabulary for discussing the stereotypes they were using and so for me, as the only "other," it felt futile, even hazardous, to confront them. In the absence of understanding the social dynamics as relations of exclusions and privilege, I criticized myself, internalizing the messages as reflections of my lack of abilities/capacities. In hindsight, I wish I had some understanding of social relations and how cultures worked. It would have made this translation so much more bearable.

So, instead of being accepted as another distinctive member of a heterogeneous community, I got the clear message that I should strip myself of cultural markers. Be like everyone else. In a culture where my differences grew by the day, I wanted to do everything to minimize these differences. I did not want to make waves, did not want to be labeled. It was either that or be the cantankerous bitch, the primitive third-worlder, the ignorant Indian, the hopelessly middle-classer, and so on. Not surprisingly, assimilation was the path I chose.

The Culture of the Classroom

With the "absorbing" attitude I have adopted the "I will work hard" resolution. I am ready, willing, all eager, and inspired to be the evolutionary biologist I have always dreamed of.

. . . I did not realize how different this system is and how much education about education was in store for me! First, I needed to learn the currency of coursework: credits, units, hours. Then it was off to classes. It is difficult to figure out notions of "authority" and "power" in the classroom in the United States. Superficially, the dynamics and the professor/student relationship seem quite egalitarian. You call the professors by their first names. Students assume an air of familiarity with the professor and the light banter is quite refreshing. At first, calling professors by their first names seemed almost blasphemous. Once I got used to that, came my cultural notions of what first names meant—friends. A couple of times I know I crossed the boundary of professor/student and I was told in no uncertain terms that this was inappropriate. Despite appearances there was not much difference in authority structures. I've begun to hear about assumptions made about South Asians and Indians. Friends have told me about professors who talk about Indians lacking the skill to write or think critically; other friends have been told that they do not "act" like scientists. An Indian woman is assumed to be passive, quiet, and obedient.

I keep wondering how I am doing. But it seems impossible without parading my insecurity, which no one else seems to do. Course performance is a secret. In India, it is common to ask your friends for their grades. Here it is secret. I feel I am left floating and therefore am never sure about how I am doing relative to others.

Classroom dynamics have been a nightmare. I feel invisible. I would desperately try to catch the professor's eye to be included. This exclusion and alienation was paralyzing, and slowly, with time, I find that I have withdrawn, first in mind and then in spirit. This social "disconnectedness" manifests itself in the classroom. People talk before and after class and I often have nothing to say. I find myself well and truly the "other"—a nonhuman amorphous being. The problem is clearly me. I am the problem.

Many of my classes involve discussions. The motto seems to be—be aggressive, opinionated, and above all: TALK! And the object of talk is to dismember, tear to shreds if possible, the papers we read. There is nothing healthy about this. It has a vicious, cannibalistic air about it. In addition, an obvious display of self-glorification and egotism surrounds each discussion. Not only are you tearing to shreds someone else's work, you score a point with each tear and anyone else in the room you refute or humiliate. I just heard that a student recently flung a paper across the room because he just did not like it! There is little room for doubt or ignorance—graduate education appears to be training in camouflage or better known as "fake it till you make it!"

The connection between the social relations within science and that of academic work was a difficult one to make. If it was difficult to work it out in my social life, it was even more difficult to make that connection in academic performance. Despite a strong ethos of egalitarianism, the academic world I was experiencing systematically reinforced traditional norms of authority and power. It was the professor who made up the exam and decided on the final grade. It was the professor who decided how I performed in class discussions. At first I did not realize how crucial it was to make a good impression. In India, unknown professors graded the final exams and each student paper had a number and not a name, but here the professor had a face and an opinion of each student while grading. Notions of gender, race, ethnicity, and nationality were necessarily an active part of this process. The assumptions by many that I did not speak English were just the tip of the iceberg.[2]

My new academic world, like the larger culture, also subscribed strongly to notions of privacy. Information about grades, performance, stipends, and salaries was considered private.[3] There was a strong secrecy surrounding performance, and it was virtually impossible for me to figure out how I was doing relative to others. This bred deep levels of insecurity that I later discovered everyone around me shared although it certainly was not apparent in daily interactions. This was deeply connected to issues of dedication. There was a strong sense that students and faculty were there because we loved science and learning. Yet, the culture was very competitive. While no one spoke about their grades, daily interactions were peppered with more subtle indicators. We were keenly aware of who got to go for lunch/coffee/beer with faculty, who was invited to do research, work, and hang out with faculty, as well as who received compliments from faculty. Reports of compliments they had received slipped into students' conversations. It was a culture that bred competitiveness and a deep insecurity. The individualizing and privatization of experience and emotions was the mechanism behind the "divide and conquer" strategy that prevented us from understanding what we had in common and made possibilities of alliances and coalitions very difficult.

The competitive, macho culture was most apparent in discussions. It was a culture of vicious criticism. The majority of the time was spent on what was wrong with papers and little time was spent on what might be interesting or useful about them. This was perhaps the single most important "skill" we were taught and expected to learn. The better we perfected this, the better we were regarded as students. What was ironic was that science was considered "unemotional." And yet, I have rarely seen more emotional or passionate people. It was not that people around me did not tolerate emotions at all, but rather they did not tolerate a particular kind of emotions—those associated with the feminine: crying, displays of insecurity, confession, self-effacement, timidity, giggling. I noticed, however, that "strong" emotions (read as masculine) were coded as revealing one's passion for the work. Faculty could yell, scream, or throw things in the labs in fury yet still be considered excellent scientists.

Class discussions did not have moderators or facilitators, only leaders. There was absolutely no focus on creating a classroom climate in which everyone felt comfortable speaking. There was no attempt to work with students who said little or nothing. Saying nothing was assumed to be an indication of having nothing to say and by extension being a poor student. There was a profound assumption of meritocracy—if students have something to say, they will talk, and what they say will be evaluated as demonstrating their quality of mind. In short, science faculty had no apparent regard for the actual skills of facilitation, teaching, and mentoring that would create a classroom dynamic in which all could participate and thrive.

Scientific Temperament

It is curious how people who initially assume that you cannot speak the language assume so much once they discover you can speak it! But I feel that no one appreciates or understands where I'm coming from. Without seeming ignorant and ill prepared, how do I explain that I had never read a scientific article or even seen the major journals in my field? How do I explain that only one of the faculty at my undergraduate university had a Ph.D. and that none had a research program? Handling expensive equipment is terrifying! Most labs in India are poorly equipped and any expensive equipment is protected. Brought up with this reverence for technology, I was taken aback when someone told me last week that I am seen as "not curious." I did not realize that my hesitation and fear of damaging expensive equipment is perceived as a lack of curiosity and interest. I'm growing to see how different my background is. I realize that growing up in Indian cities all my life, images of nature largely lie in the alien realm of David Attenborough and Jacques Cousteau. I have never had close access to forests. I see that most of my peers have grown up exploring the woods, hiking, visiting national parks or the stream behind their house. Many are naturalists or aspire to be. I do not share this socialization and it feels ominous not to know a single species of plant or animal around. Not a great start for a budding ecologist.

I met with several universalist assumptions about science and scientific ability that interfered with my learning. In my undergraduate education, I had never read a scientific journal or seen many research labs. Even large Indian universities do not have many resources compared to institutions in the United States. None of the professors in my college had a research agenda (most did not have Ph.D.s). There was very little technology, and what was available was fervently protected. The sole copy machine at the college was housed in its own room in the library with a paid operator who controlled access. In the United States, in contrast, the ethos of science was one of play. It was routine to see undergraduates playing with equipment (and often breaking it) and they were provided with expensive chemicals and restriction enzymes that many research labs in India could not afford for the most experienced scientists. It has now become apparent to me that ideas about what constitutes a scientific temperament and "good" scientific practice are very culture-/nation-specific, deeply influenced by the particularities of historical, economic, and political contexts. I wish I had known.

Initiations

POPGEN (short for population genetics), an informal discussion group, meets every Thursday night. It involves a gathering of professors and graduate students. It was started in the "good old days," we are told, in the true masculinist tradition of talking shop over beer into the early morning hours (while, of course, the "girls" cooked and cared for the babies). I know all the people and I like most of them, but put them all together in a room!

The session traditionally starts with announcements and then the speaker is introduced; the object here is to be witty and male. The degree to which this is accomplished depends on the status of the introducee (just how much of an "old boy" s/he is) and the introducer (just how steeped in tradition s/he is). On a really good night you might hear something like, "What I really like about X is that he always gets his priorities right. When not working, it's coffee, beer, and sex and when he's working it's much of the same." Loud chuckles, laughs. Then the speaker starts talking and can be interrupted anytime, and then it is a free for all. The culture is violent, insular, familiar, steeped in inside jokes. An informal history is passed down the generations of all the ones POPGEN "got" and the "you won't believe what he said . . ." quotes. The ones POPGEN "got" involve the historic sessions, the speaker usually being an outsider.

The sessions also serve as an "initiation" rite for students. Anyone in the field "has" to present their work as an unwritten rule. So you prepare in dread and everyone looks at you with empathy as you pass the halls the week before, quite akin to the sacrificial lamb being led to slaughter. With time you are the seasoned veteran with a clique to sit with, nudge, crack jokes in the back, and incite laughter. Most sessions are, however, fairly cordial with an undertone now and then that clearly reminds you that the "old boys" are alive and kicking.

POPGEN was one of the more fun and also defining experiences of my graduate school days. It was a space where you built community, where everyone in the field gathered together one evening a week to informally discuss science. Everyone who worked in the area of population genetics was there. Most of the faculty also made it a point to attend. Very often we got people from neighboring institutions as well as our own. These sessions were intellectually engaged and exciting, and I certainly learned a great deal from them. It was a ritual I attended every Thursday night, and I was initiated into it as well. POPGEN was the in-group, the defining collection of individuals, where multiple lab groups came together. The idea of informal but engaged intellectual exchange was wonderful with tremendous potential. The spirit of engagement, collegiality, and informality was philosophically wonderful, but the ethos betrayed its cultural assumptions—sometimes in clear and terrifying terms. Underlying notions of machismo, power, and masculinity pervaded the culture as a consistent theme in subtle and unsubtle comments, actions, and conventions.

Sharon Traweek in her work on high energy physics suggests that the culture of science is the "culture of no culture" because scientists claim not to have a culture. POPGEN was a moment where I was most struck by this. The culture was obvious and well defined. POPGEN had a clear ethos. It was about exhibiting one's intellectual/scientific rigor through a display of aggression, competition, machismo, and criticism. It was clear what was deemed funny and witty. Cliques were full of innuendo and history. Being "nice" was frowned upon. The initiation ritual was clearly about learning to face the "firing squad" (a word often used). This rigor was proportional to the degree to which one could tear apart others' arguments, humiliate them to the point that they were speechless or in tears. The necessary response was also clear—fight back. Breaking down into tears was a sure way to display your scientific unworthiness, of just not being "cut out" for the profession. It was ultimately about being one of the "boys." In retrospect, what POPGEN represented for me was the universalization of very particular cultural norms: American-ness, whiteness, maleness, heterosexuality, argumentation versus dialogue, along with the assumptions that these norms grounded good and rigorous science.

Dispersions of Identity

Despite all my emulations of furniture and attempts to blend in, I did not succeed. Rather than blend in, my identities seemed to take on an increasingly significant role. Like rays of light dispersing into colors of vibrant spectra on passing through a prism, my travels through the prism of objectivity in halls of science have dispersed into a dizzying and fragmented array of multiple identities. Never in my life had my identities seemed so pregnant with meaning and symbolism, defining and determining as it were issues of professional life and death.

The "prism" has been very useful in thinking through my experiences. Coming to this country, I did not affiliate with any "identity." The culture of the United States, academe, graduate education, and the sciences nonetheless all proved to be prism-like—dispersing and fragmenting the "I." The rest of this chapter tracks the process of my growing political consciousness. To be sure, now I know that we all have multiple identities, belonging to particular or multiple genders, races, ethnicities, classes, and nations at the same time. Yet when no one around me shared these multiple identities, my growing realization of the importance of identity came in fragmented ways—coming to identify the individual colors of a prism. I explore my growing realization of my multiple identities. What I hope these narratives show is that identities "no matter how strategically deployed are not always chosen, but in fact are constituted by relations of power always historically determined" (Viswesaran 1994: 6).

Gender

The silences of insecurity are growing deafening and I've begun talking to friends about the deep sense of insecurity I feel. And to my astonishment, I am not alone! Some of us have decided to start a discussion group on "Women in Science." It has grown into a format of women bringing their experiential accounts and the group attempting an analysis. To our surprise some men have joined. Is this experience gender specific? Don't we all feel insecure and as imposters? So, in the true scientific tradition three of us decided to test the hypothesis. Becky Dunn, Lynn Broaddus, and I devised and administered a short questionnaire on the effects of graduate school—on self-confidence, ranking of self, perceived perception of peers, considering quitting, strengths, weaknesses, future goals, and so on. We were quite astonished by the striking results. Gender differences seemed to be consistent and at times remarkable. And so, we presented our results in an open seminar to the two departments (Subramaniam, Dunn, and Broaddus 1992). The room was packed. The response was hardly surprising. Our study was interesting! The audience was deeply reluctant to accept the underlying suggestion that institutional structures discriminated along lines of gender. There was far too little diversity in the departments to include race, class, or sexual orientation. "Had we considered age? That might explain it"; "The problem was not so much women's lack of self-confidence but overconfidence of men"; "Why were we worried about this? At least one-fourth of the students had expressed aspirations of going to a large research-oriented university like ours. Surely we did not want more!" The comments ranged from questioning, to accepting, to legitimizing our data. The results and process were, however, very empowering for many of us. First of all, our perceived insufficiencies and doubts were no longer individual but formed a collective. A pattern had been documented and comments like "Oh! that happens elsewhere, but not here!" were not valid anymore. With the survey, and, above all with the individual work of women in the two departments, things began to change a little. If nothing else we felt legitimized in treating it as an issue.

Our analysis grew. For example, we observed faculty searches. Graduate students have a representative on the search committee and also meet with the applicant as a group. Graduate student opinions are presented to the faculty, who then vote. Various individuals' impressions of the candidates of course float through the grapevine. Women are traditionally either the "mouse" ("she's much too quiet," "she has no opinions") to the proverbial bitch ("she is always dying to pick a fight," "she's too argumentative," and on one occasion "she wore pants to the interview!"). We've had "nice country girls" to "she's really nice and I really like her, but . . ."

What are the criteria of the process, we wondered? Superficially it would seem that objective criteria exist—of papers published, seminars presented, ability to relate, broadness of interests in biology. Once in a while a candidate appears who fits all the above criteria and who also seems to be a radical and you are thrilled. Clearly the candidate is best qualified. But as we listened more carefully we heard that his or her thinking was "diffuse," "fuzzy." We asked what that meant and usually got a shrug and an "I don't know quite how to put it, well . . ." With time, we realized that the criteria were changing with the searches. Suddenly we were marginalized for having an "objective, non-discriminating" voice. Ultimately, the candidate that was "best" suited was someone who would be a good colleague and the "system" would replicate itself, and thus, the written and unwritten rules and norms of academia were legitimized and perpetuated.

I still remember how powerful this moment was. On the one hand the reaction was depressing. There were some whom we could never convince, never mind what data we produced. There were many who refused to believe that the data captured something about scientific culture. It was a powerful moment for many of us, however, as it led to changes in departmental practices. There was increased representation of students on faculty searches. And suddenly faculty candidates were asked about their perspectives on women in science. For a little while at least the issue was alive.

In retrospect it was entirely a project about culture and climate. What was striking about the results was that all students were having a rough time. All experienced a severe reduction in self-confidence and self-worth, although the decrease was much more dramatic and significant among the women. Even our own understandings were very much within the framework of the sciences. There were some "bad apples" who were making the life of students miserable. Women needed to be better represented among the faculty and in search committees for future faculty. Students needed better advice and faculty needed to be held accountable. None of us knew about the feminist studies of science. We did not recognize that our work had a history with which we could connect our present struggles.

This is where my future would lie. As the feminist studies of science has repeatedly argued, addressing the perspectives of those underrepresented is not a special interest issue but renders visible privileging structures that had been previously normalized—and which impact *everyone*. I am reminded of well-publicized anecdotes such as one from the neurobiologist Ben Barres who after transitioning from a female to a male, from Barbara Barres to Ben Barres, recalls hearing a male colleague praising his work as "Ben Barres" while simultaneously disparaging the work of his sister, "Barbara"; these are powerful reminders of the climate for women in science (Barres 2006, Baty 2010, Begley 2006). Attending to how scientific cultures function to normalize some bodies and behaviors while rendering others aberrant impacts all scientists, men and women, especially future scientists who desire something other than the prototypical disembodied, a-cultural fiction of scientific rationality.

Race/Ethnicity

Resolving my Indian and third world identity was more difficult and it took longer. It was clear how white masculinity was the normative mode of engagement. Most department members were from the United States and white. Whiteness thus became an unmarked, neutral category for science. Since there was little diversity in the department, I tried my level best to ignore that it was a problem until it was so blatant that it stared me in the face. As with gender, the most obvious moments came from the occasional faculty of color who came for talks or presentations. The stereotypes that often surfaced in the department shocked me. In one case, for example, a candidate was introduced for his job by detailing all the violence he had encountered because of his brown skin and presumed nationalities! Candidates are usually introduced with a history of their scientific accomplishments. I could see how innuendos emerged and how they slowly grew to become facts of sorts. At times with male scientists of color, the old trope of the oversexualized male of color surfaced. Watching how social innuendo translated into claims of scientific unsuitability was fascinating. In many cases this felt personal, and I felt hypervisible as a nonwhite, non-U.S. student. But it was these rare moments that gave some of us a window into the racialized world of science. After all, I came to realize that science was inhabited by scientists and that the same prejudice and biases that pervade mainstream culture are likely to suffuse the hallways of science.

In most of these cases, I did not confront my colleagues, and they seemed oblivious enough. And I was not alone since some white colleagues also saw the process for what it was. It was through this process that other memories came flooding through. I had thought them curious and filed them away. I had felt uneasy about the comments but had never confronted the racism, and they all rushed into my consciousness. The professor who said, "She came with a stone on her forehead and look what we've done to her"; a fellow student who read an article somewhere about India where Indians during some religious festival stood in line and jumped down into a river below and drowned. Was it true? Volunteering for pro-choice events, the rationale was often "this sort of thing (illegal abortions) happens in third world countries, but we are part of the industrialized west, it is unacceptable here . . ." It was coming back, all fitting together.

There were no other Indians in the department or indeed in women's studies and not many other third-worlders or students of color in either location. My Indian and international friends were mostly in other fields, often in other universities. North Carolina is particularly stark when it comes to race. All janitors in my department were African American and all faculty but one were white. Similar patterns confronted me in stores and fast-food restaurants. This lack of diversity did give me pause, although it was never personalized. Still, I had become a racialized subject.

Observing the reproduction of racialized subjects of science, I came to understand what was ultimately an act of translation: I was learning to encode cultural difference within the language of meritocracy. A racialized subject is never a simple category. I saw that the African American candidate for a job is read very differently than the Middle Eastern candidate or the South Asian one. Men and women are racialized very differently. Someone like me who grew up middle class is read very differently than someone who might have grown up poor in the United States. Thus different histories of race, I observed, were salient within my scientific culture even though they were seemingly unnamed. Unacceptable or divergent cultural manifestations were reworked as deficiencies of scientific ability or practice. Ironically, it was these same assumptions of race and national origins that helped me pursue my interests in women's studies. Taking courses outside the biological sciences was frowned upon (especially if it was women's studies). When I decided to start taking courses in women's studies, my interest was supported because some faculty felt that they ought to foster and support my interest in "feminism." After all I was from the third world and in need of such influence and education. Indeed, some of my white U.S. counterparts were not so lucky and were asked to put more time in the laboratory instead of "wasting" time in irrelevant courses outside their area of research.

Sexuality

I was entirely oblivious to the heteronormative culture of my graduate training. It was only when I began working more substantively in women's studies that I recognized that assumptions about sexuality pervaded the culture. I still remember the moment when this hit me. I was completing my doctoral work in evolutionary biology while beginning a position in a neighboring institution where I had been hired to establish a "women in science" program. Every morning, just before sunrise, I would stop by the greenhouses to collect data on my plants and subsequently head to my women's studies job. In the evening, I returned to the greenhouse and my office in biology before driving home. One day while en route from the greenhouse to my job I caught myself in the midst of a transformation. I realized that I had begun responding to my bifurcated life by quite unconsciously transforming my persona in the sciences into my persona in women's studies. Earrings went on, plus a waistcoat for color and to cover up the dirt stains, to sanitize my mediations with the natural world. When I drove back from women's studies to the greenhouses, I realized that I enacted the reverse transformation. In the sciences, I had learned to strip myself of all markers of my sex or sexuality. Not only could I not be marked "female," but I also needed to appear to be oblivious about my appearance—a practiced casual grunge. In field biology, clean clothes are the surest sign of not working! The two worlds required phenotypic transformations that required deliberate and practiced performances. Like gender, disciplinarity should also be understood as a performance.

One of the attractions of the sciences was precisely the absence of pressure in having to look "nice." But I was surprised by the unofficial rule and unarticulated pressure to not look "nice." For field biologists, sex and sexuality were very important. What you wore, how you carried yourself, whether and how you dealt with your sexuality (with peers and faculty) were filled with meaning and had profound consequences from comments in the hallways to being "advised" on propriety. Any visible articles of femininity usually elicited numerous comments about one's appearance. I suppose I had enough battles to fight that I never took this one on. In hindsight I remember that I stripped myself of all markers of my feminine self. I stopped wearing earrings, gradually threw away dresses or any clothes that could be construed as feminine, started cutting my hair short. It is rather extraordinary to me to realize the power that climate had over me; in fact, I had little awareness of what I was doing. It is only the power of hindsight and analysis that allows me to recognize the profound impact of the culture on my day-to-day existence. This fit between a scientific identity and a masculine one has profound implications for the training of graduate women since it suggests that being at once a scientist and a woman is not possible (E. Keller 1987). With its roots in a Christian clerical tradition, science emerges from an "exclusively male—and celibate, homosocial, and misogynous—culture, all the more so because a great many of its early practitioners belonged to the ascetic mendicant orders" (Noble 1992: 163). Like the clerical coat, the lab coat signified devotion and dedication. In my case, I made a deliberate choice, informed by my feminist consciousness, to perform my appropriate and expected gender role, in order to avoid the consequences of "gender trouble" (Butler 2006) and any ensuing disciplinary trouble! The rules of dress, however, were entirely different at social gatherings, parties, and receptions. Here, assumptions of compulsory heterosexuality and American femininity were desirable. Thus there were expectations of masculinity within the walls of science and of femininity outside those bounds (E. Keller 1987). To remain a whole and coherent individual required phenotypic transformations or a deliberate feminist consciousness where one deliberately performed one's appropriate and expected gendered roles or battled the consequences of one's "gender trouble" (Butler 2006).

Dissembling

GAMES ACADEMICS PLAY ✗ *so true*

And so I have begun to explore exactly what about the environment around me nurtured my baseline paranoia—that because I was different, everyone was out to get me. What about the culture made me insecure? I began to observe behaviors, attitudes, underlying presumptions of people's comments, actions, and beliefs.

> The "You don't work as hard as me type": This is an extremely insidious type. They constantly have to let you know how late they stayed up the previous night. "God! you're taking time off? Lucky you." It almost seems like an unwritten rule that if you are a graduate student you must have a tired, harried, longsuffering look about you! Being visible around the department is crucial and students develop strategies of advertising their presence.
>
> The I am buddy/buddy with faculty type: These types are constantly reinforcing their faculty connections. "Oh, I went for a beer with . . ." or "While I was chatting with . . ." The implication is that they are fraternizing with "the powers that be" and are on their way up.
>
> Things are going fantastically well type: These are the perpetual optimists constantly stating how well they are doing irrespective of their actual performance. Often the stories through the grapevine tell you otherwise.
>
> The "It is your fault" type: They espouse a singular dedication to the system. If you present them with a critique or criticism of the system, the fault is yours—"You asked for it," "You are misinterpreting." They make up more tasks for you. These are lethal.
>
> The "You're so emotional" type: This is a perpetual problem. "Oh, he did not mean that . . . ," "You're overreacting," "Chill out." The basic premise is that you are feminine, and therefore unqualified.

Other notions of gender and the tightly knit history of masculinity and objectivity also invaded. In a class for which I was a teaching assistant, the professor misspelled a word and knew it was wrong and stood staring. I corrected him. He laughed off his ignorance as "I cannot spell, therefore I became a scientist!" thereby implying that because I could, I should not be one?

You may or may not find this typology useful or true to your experience, but I am sounding a warning. I believe that the socializing and the game playing are an active part of maintaining the "old boy network." Most significantly, what

I am suggesting is that the acceptance of these value systems will not get you into the network. Instead, the value system is designed to exclude the ones who cannot stay as well as to condition the ones who can. For example, spending all your time in the department is valued not because you are getting more work done, but because you are more easily accepted as "serious" or "committed." This structure is never innocent—making it easier to exclude those who cannot stay around the clock and who may even work more productively because of their lives and commitments outside the labs.

Culture and Politics of Science

Coming to political consciousness profoundly transformed my perception of the world around me. A world that once seemed about producing "good science" suddenly also became a world about producing "good scientists." I came to learn that graduate education was about "making" scientists and reproducing a set of cultural practices and behaviors about good scientists. As I observed the world around me, the expectations and norms of the culture of science emerged. It reminds me of Toni Morrison's wonderful observation on learning about racism in American life: "it's as if I had been looking at the fish and suddenly I saw the fishbowl" (Morrison 1992:17). Indeed, the tools of women's studies enabled me to understand science as a historical and social set of structures, practices, and epistemologies of knowledge. Graduate school for me was a place of policing these boundaries—the weeding out of those who did not participate in these cultural practices. And the clashing of cultures of a middle-class, Indian, postcolonial feminist and those of western science was precisely what I encountered—the disjunctures and dissonances of these cultural practices. My love of science was grounded in a fascination of understanding the natural world around me. But what I failed to recognize was that all knowledge is produced by a set of scientific practices generated, agreed upon, and regulated by a community of scientists. The crucial step for me was the recognition that "good" scientific practices were less about good science than about reifying historically generated and reproduced practices.

While learning and doing science, graduate school was also teaching me the set of cultural practices as an important (but unannounced) part of science. And these cultural practices were hardly universal. They were (and still are) a set of practices bound by the historical roots of science in western, white, male, heterosexual culture. And this was precisely what I realized the good students were so good at—aligning themselves with the markers of "good" scientists until it was soon almost naturalized, subsumed, and ingrained. These norms and rules are hardly totalizing, however. Students and faculty resisted them in all kinds of ways—not all of the rules all of the time but rather some of the rules some of the time. Figuring out the outer bounds of acceptance was crucial. And these bounds varied depending on the race, class, gender, nationality, and sexuality of the individual. It was about negotiating a complicated set of social relations—playing it well enough to remain credible and little enough to stay sane!

And so I came to recognize these scientific rituals and postures—when they had to be practiced and when not. I came to recognize when people were trying to put me down, pull rank, appear smarter, point to my insufficiencies, or blame me as the problem. This new perspective created for me a space where I began to enjoy "doing science." Ironically, it took a political consciousness to begin to enjoy doing science again. I had at this point in some ways traveled a full circle—returning to the fascination of science I once felt, although it was more complicated and nuanced. Identities, cultures, and expectations began to fall into place. And it showed. Students began to come to me for advice. How was it, they asked, that I was doing so well? It felt so good!

As I think back on my graduate education, I wonder what it might have been like if women's studies had been an interdisciplinary option of study instead of a crisis management strategy. What if we could break this cycle of silence? Science and the prism of objectivity split my individual self into an awareness of the dispersions of my identities. The tools of feminist scholarship allowed me to refocus these multiple identities to feel whole again. Interdisciplinarity has enriched my intellectual development more than I could have ever imagined. This is the stuff dreams are made of!

CHAPTER EIGHT

Resistance Is Futile!
You Will Be Assimilated

Gender and the Making of Scientists

> What was so new about these projects of docility . . . ?
> [An] uninterrupted, constant coercion, supervising the
> process of the activity rather than its result and it is
> exercised according to a codification that partitions as
> closely as possible time, space, movement.
>
> —Michel Foucault, *Discipline and Punish*

> Making visible the experiences of a different group
> exposes the existence of repressive mechanisms, but not
> their workings or logic; we know that difference exists, but
> we don't understand it as relationally constructed. For that
> we need to attend to the historical processes that, through
> discourse, position subjects and produce their experience.
> It is not individuals who have experience, but subjects who
> are constituted through experience.
>
> —Joan Scott, *The Evidence of Experience*

During the hey day of the television series *Star Trek: The Next Generation*, I remember being transfixed by the episodes on the Borg.[1] Their repeated and relentless "Resistance is futile; you will be assimilated" was a catchy and unforgettable slogan. Steeped in a project on the culture of science, I was struck by the resonance of what I was hearing from the Borg and from graduate students in the sciences about the culture of science. The singular focus, the complete dedication, the all-absorbing culture, the strict adherence to rules and order, the intolerance of deviance of any kind—these were some of the broad descriptions that emerged about the sciences. The Borg shared the same passions. I use the title not as a way to embrace the antitechnology or anticollective action implied in the Borg episodes, but instead to refer to assimilation in its most literal sense—about the requirement for compliance to gender norms and rules in the culture of science, to a set of norms and rules that refuse to go away despite the entry of women in significant numbers into the hallways of science.

Over the past two decades, the social and especially the feminist studies of science have developed a provocative theoretical framework in which to locate the processes by which cultural constructions of gender are related to the creation of scientific knowledge. Sharon Traweek's path-breaking work reveals that like all institutions, science also has a "culture," albeit a "culture of no culture"! Scientific practice, she notes, requires an objective, rational, asocial, decontextualized researcher, and a person immune from context, from culture (Traweek 1992). David Noble provides a sustained analysis of how the history of science has developed and shaped a culture that has systematically excluded and marginalized women and people of color. He points out that scientific culture originated within the western Christian clerical tradition. The history of science is a history where women and people of color have been systematically excluded and/or marginalized. These histories have consequences and have shaped the contemporary culture of science. In tracing the roots of western science, he observes that

> several habits and characteristics of modern science have often been noted: the strict separation of subject and object, the priority of the objective over the subjective, the depersonalized and seemingly disembodied discourse, the elevation of the abstract over the concrete, the asocial self-identity of the scientists, the total commitment to the calling, the fundamental incompatibility between scientific career and family life, and of course, the alienation from and dread of women. (1992: 281–82)

The belief that science does not have a culture, the resoluteness with which the subjective and experiential are relegated to the personal, are ultimately intellectual and epistemological moves. Science need not attend to the experiences because it claims a "depersonalized and disembodied discourse." Scientists' individual identity does not matter because scientists are interchangeable, all independent nodes in the production of knowledge.

Reflecting on the consequences of science's construction of "nature as female" and "mind as male," Evelyn Fox Keller points out that "this poses a critical problem of identity: any scientist who is not a man walks a path bounded on one side by inauthenticity and on the other by subversion" (Keller 1985: 174). These works suggest the importance of a more sustained analysis of how contemporary scientific culture functions and how variables such as gender, race, class, and sexuality shape this culture.

This fit between a scientific identity and a masculine one has profound implications for the training of graduate women since it suggests that being at once a scientist and a woman is not possible (Keller 1987). A woman would have to be trained in such a way as to render her gender identity invisible or always conflicted and difficult. Appropriate and adequate mentoring in this context would have to make this a necessary component of developing a professional identity. Therefore, mentoring, rather than helping students grow into a new sense of self, is directed at requiring women students to give up a core and familiar sense of self—what Mary Wyer and I have called a (de)mentoring because it removes rather than contributes (Subramaniam and Wyer 1998). This is to suggest that women need to be "untrained" as women and "retrained" as scientists. How do faculty in the sciences achieve this untraining of women and retraining as scientists? How can feminists intervene in this process? What new visions of mentoring can we imagine that do not render "woman" and "science" as incompatible identities?

Armed with these insights, we ventured into science land on an NSF-funded project to open conversations between faculty and students in the sciences about the culture of science and the mentoring of women graduate students. In the absence of a well-developed research base on women in graduate education, we hoped to begin formulating a theoretically informed exploration of graduate women's experiences, one that could explain their underrepresentation across most science and engineering disciplines. Again, as with the previous chapter, some of these observations are likely generalizable to graduate education across fields.

The project I describe was based on an innovative project and methodology developed by Mary Wyer for a previous NSF project (Wyer 1993). One of our first discoveries was that graduate education is structured less around the classroom and more around a model of apprenticeship, that is, a protégé-master model. In such a one-on-one model, interpersonal communication and relationships are critical, and social markers of gender, class, ethnicity, and sexuality ubiquitous. Yet, as Traweek suggests, talking about interpersonal communication, relationships, and social markers is forbidden. At the same time, what the literature on mentoring reveals is the implicit kinship basis of the model—the word *mentor* comes from the *Odyssey*, where Mentor stepped in to guide Telemachus when his father (Odysseus) was away. So the problem of protégé-master is that it is a reproductive model, reproducing the master in the protégé who cannot handle difference; resistance is futile; the mentee's only hope of success is to adopt the master's identity.

We also came to realize that graduate education is a unique training ground in the educational ladder. One enters graduate school as a "student" clearly subordinate to the faculty and in search of training from them. A few years later, however, one is expected to leave as a "colleague" to the very same faculty. While during undergraduate years we learn about science and might perhaps even learn how to do experiments and interpret data, it is during the graduate school years that students learn how to "be" "scientists." For this, they must learn to present themselves as credible professionals—how to design and carry out research projects, choose interesting and productive research topics, plan and carry out the logistics of experiments, give talks, network, discuss science with colleagues, procure grants, publish results, recruit and motivate good students. Graduate school is thus a critical phase in the "enculturation" of scientists and a place for the reproduction of a particular professional "identity." The enculturation of scientists in our project ended up being the production of knowledge makers, who value *not talking about* and *not recognizing* the social world they create, maintain, and reproduce (Subramaniam and Wyer 1998). How does this culture function? How does it reinscribe particular notions of gender, race, and class with the next generation of aspiring scientists?

[handwritten margin note: How does this impact the type of science that gets produced?]

In order to establish a dialogue between faculty and students, we began with facilitating a conversation between faculty (men and women) and women graduate students about the strengths and limitations of graduate education for women. We worked with four departments (Chemistry, Molecular and Cellular Biology, Ecology and Evolutionary Biology, and Mathematics) at a large southwestern public university. We chose these departments because they had supportive chairs and because they represented different organizational environments for doing research (Fox 2000). We were particularly interested in identifying distinct faculty and student concerns around issues of gender in graduate education.

Following Wyer's (1993) method, we developed four groups: two groups of students and faculty each, in order to replicate our findings. Two groups were composed of women graduate students, with ten students in each group. The other two groups were composed of ten men and women faculty. Groups, like individuals, can often develop idiosyncratic behaviors; two groups therefore made the project more robust, within the limitations of the money and time available to the personnel of the project. There were several criteria for choosing students for this project among the many in the department. These included creating a diverse group (across ethnicity), year at graduate school, experiences in graduate school thus far, kind of research, and small and large lab groups. We chose students who were open-minded, articulate, interested, and committed to being part of the project. Faculty were chosen to create a diverse group along the same lines as students, for their reputation of being supportive of women graduate students, their commitment to seeing an increase in the recruitment and retention of women in science, and their commitment to improving graduate education in general. Two facilitators facilitated five sets of meetings with the four groups, for a total of twenty two-hour sessions.[2] All sessions were taped and subsequently transcribed.

A central concern of the research was to attend to the power inequities between faculty and students. The faculty and students came from the same department, and in a few cases faculty members in the project served as dissertation advisors to some students in the student group. It was clear that this would not produce an open and honest dialogue if students or faculty knew their advisor or student was participating. An innovative aspect of Wyer's research design was to create a dialogue between faculty and students through facilitators without them knowing each other's identity (Wyer 1993). The student and faculty groups met independently and heard about the responses through the work of two facilitators without meeting each other. Therefore, the identity of participants in any group remained anonymous to the other groups.

The Encounter, the Silences:
Some General Observations

We began the discussions by asking graduate women to describe their experiences, their interactions with faculty, and their departmental cultures. We encouraged participants to share anecdotes about their experiences with graduate education. The experiences women graduate students shared revealed a great deal about the place of graduate students in the larger fabric of scientific culture and the frustrations and conflicting messages they received from faculty.

What was striking when talking to the two groups was their stark ignorance about each other. It seemed to us that there must be profound silences that accompany their daily lives where although they inhabit the same world, they rarely talk to each other about their daily experiences. It seemed as though faculty and students came from different worlds even though many shared the same laboratory. The issues that were persistent in student experiences were the lack of and the need for greater communication between faculty and students. While there was departmental variation, on the whole, students felt there were not enough occasions for faculty-student interactions and that overall they did not believe faculty cared.

Given that graduate education is largely structured around a one-on-one mentoring system, individual interactions with graduate advisors take on critical significance. In our project, gender and ethnic identities became very important in individual interactions and in departmental cultures. While there were a few examples of blatant objectification of women, in seminars and lectures, these were not prevalent in these departments.

The bulk of examples from students were subtler. Most spoke about small, daily interactions with faculty. For example, a student recounted a committee member who returned a proposal to her telling her basically, "It sucks. You'd better start from the beginning." The student described herself as being deeply depressed for weeks, unable to look at the proposal she had put a lot of work into. She then found the professor to discuss the proposal, and during the conversation it emerged that he thought that apart from the introduction, the rest of the proposal looked fine. She remarked, "He got so pissed off reading the introduction that he put a horrible comment at the beginning and returned it." She pointed out that had his comments been more precise, it would have saved her a lot of anxiety and dejection.

What seemed significant to us as facilitators of these meetings was the disjunction between the two groups—students' conviction that their anecdotes were real and very common, and faculty's insistence that students were over-interpreting or overreacting or just plain wrong. Students insisted that gender and ethnicity played a salient role in student-faculty interactions. While there was variation among and between men and women faculty, it was striking that faculty as a group held more similar views with each other than women faculty and graduate students. It was remarkable that faculty, who were once graduate students, could remember so little of their graduate student experiences. Yet, it is important to note that women faculty did recognize their gendered experiences as women. Some attempted to educate their women students, to soften the harshness of the expectations, while others worked hard to "toughen" women students for what lay ahead. It was apparent that the culture of science enabled a deep conformity, so it was not obvious that women graduate students always found more supportive climates in laboratories run by women.

LABORATORY CULTURE

For disciplines such as chemistry and biology, the laboratory proved to be a crucial location where the negotiation of social relations reigned supreme. Here, graduate students were confronted with interacting with the same men and women day after day. Each lab group seemed to have its own set of issues, often dependent on the lab advisor. Some lab advisors were very involved with dynamics in the lab, often setting rules for research and conduct—such as lab meetings and cleanup protocols. Other lab directors were absent, letting students and post-docs negotiate their ways around the lab. Often this meant that students had to take on the role of developing sign-up sheets and organizing the labs and cleanups. These advisors often discouraged students from coming to them to ask for mediation of a dispute. In such contexts women students described dynamics where they took on the role of mediators of the lab. For example, one woman in an all-male lab discussed how her advisor expected her to be the mediator and confidante in the lab, thus fulfilling the traditional "female" role. The advisor would insist on telling her all about his personal life and problems. The man would not discuss anything personal with the male graduate students. Instead, he would pump the woman for information on her peers. Where had this male graduate student gone? How long would he be gone? The woman had to play the role of mediator and interpreter.

SELF-CONFIDENCE

Another important theme that emerged was the issue of self-confidence and self-esteem. Students described their perception that graduate school was very hard on student self-confidence. Students acknowledged that graduate school was really different than undergraduate life. They felt that it took a lot of time and effort to figure out exactly what was expected of them in graduate school and the fact that these expectations varied within the faculty. They insisted that it ought to be the graduate mentor's role to teach students the ropes, explain to them exactly what they were expected to do and when, and spell these expectations out clearly at the start. Students described eloquently how hard the transition from undergraduate to graduate student was, especially for those coming from small colleges.

Students described, sometimes in detail, their struggles with figuring out how to do research, how to think, how to organize knowledge, and whether and whom to ask for advice. In a culture that privileges being knowledgeable, they felt they had to pretend, remain silent, or fudge on questions raised. "Fake it till you make it," as it is commonly referred to. Many recounted their growing appreciation of posturing in scientific culture and learning how to posture themselves.

I don't talk to many of my committee members about my future plans. If they ask me what my plans are, I tell them, "I want to be a professor just like

186

you" because that's the answer I think they want! . . . I'm not sure what would happen if I were honest. Although I think that I would get less respect from some of my committee members, as if I were not a serious student—which of course I am.

But students identified posturing as a problem systemic in the culture of science and academia. They described the many times faculty would pretend to know something that they clearly did not, unconsciously modeling ways to posture. In summary, students described graduate school as a phase that they knew little about when they entered it. The early years of graduate school were spent figuring out the system and what was expected of them—slowly muddling their way through it. There seemed few departmental or university-wide resources that could help them through it. Student experiences were often idiosyncratic, dependent on the advisor/lab director's investment in the wellbeing of students. There seemed a huge variance in the quality of graduate mentoring. In laboratory situations, students' experiences with lab cultures depended on the laboratory directors and the degree to which they were willing to create a supportive climate for women students and intervene when necessary. Through the discussions of these anecdotes students were calling for more sustained departmental and university-wide support structures for graduate students and recourse for students who were working with unsupportive advisors or labs. Also, they felt faculty should be held accountable for their mentoring practices, and that institutions should reward good mentoring.

In this project, students were not asking for blanket approval or encouragement when it was not deserved. Students stressed that faculty who were not honest in their feedback and criticism were not good mentors either. Rather, they were asking that graduate orientation and socializing become a more uniform and explicit part of the curriculum.

FACULTY MENTORING

My colleagues and I talked to faculty about their part in the graduate school relationship and asked for their views on mentoring. We asked them to describe for us what they considered good mentoring, how they trained their students, what their expectations of students were, and how they had learned to mentor students. A point that all faculty underscored was that they did not receive any training in mentoring to equip them for the job at hand, and they tried to do the best they could.

> I don't know about you all, but going and getting a Ph.D. and going into industry and post-docing for a few years after that never prepared me to teach

undergraduate classes of eighty people and serve as a mentor to graduate students. I mean, just because you have a Ph.D. in science doesn't provide any training at all for the job. There can't be any other profession in the world where you receive less training for what you actually do.

Faculty seemed hard pressed to name particular models that they used except to generally state that they gave students feedback, encouraged them, and did the best job they could.

The faculty narratives speak to the challenge of both encouraging and nurturing an individual while at the same time being the "disciplinarian." Mentoring models in faculty accounts seemed deeply impoverished. Faculty received little training, and these issues were seldom discussed in the departments in our project. In response, faculty tried to cope the best way they could—by following their instincts and experiences from their own lives, such as with children or their own graduate experience. One of the faculty remarked that many of the "jerks" in academia were the product of their advisors being "jerks." The only overt mentoring model faculty named was mentoring as "parenting"—a very problematic model for education.

Graduate students responded to faculty participants' admission of lack of training with little sympathy! They insisted that faculty were hiding behind these admissions to dodge responsibility, and that good mentoring could be learned and developed over the years, if a faculty member cared. But both faculty and students felt strongly that academic culture does not value mentoring. It carries little weight during promotion and tenure reviews, and there were few rewards to faculty who spent a great deal of time with students, often mentoring students that were not their own. Both groups felt that these were often women faculty. Students in each department could readily name these faculty members.

There were two main overall differences between faculty and students in their discussion of graduate education. First, faculty and graduate students used different languages to characterize mentoring relationships and graduate school experiences. Students stressed commonalties among graduate women's experience. They tended to have talked with others about graduate education, and to share with each other a language for it. Indeed, by the end of the second meeting, both student groups had developed a coherent and cogent narrative about gender and graduate education and about the process of graduate education in general. While there was some variation in student experiences, often shaped by lab groups and departments, students seemed to be able to agree on a general framework for graduate education remarkably quickly. Students were often astonished at how similar some experiences were across departments.

Discussion sessions provided a process to organize individual and seemingly idiosyncratic experiences of students into a collective analysis of graduate education. Student analyses characterized graduate education as a self-replicating and well-reinforced system.

Faculty, in contrast, generally viewed their relationships with students as individual, idiosyncratic, and particular. Anecdotes that students had offered in their meetings with each other as symptomatic of larger currents within graduate education, when relayed at faculty meetings, were usually analyzed as reflective of problems of individuals. Faculty did not seem to share a language with each other about the nature of graduate education, and when they analyzed difficulties they tended to use psychological terms like *self-esteem*, *self-motivation*, and *self-direction*. In contrast, graduate students preferred an idiom of culture, politics, and power. For faculty, it seemed less true that they saw themselves participating in a system; rather, they tended to see individual relationships and problems. Perhaps this is in large part because of an overriding importance of signaling expertise areas—where faculty habitually distinguished themselves from one another by different areas of expertise, and did not appreciate that they shared in a common enterprise. For graduate students, on the other hand, the process of being a protégé temporarily trumped field or expertise differences and put them in solidarity with others. Moreover, while graduate student narratives themselves seemed quite systematized, and were coherent and cogent, faculty understandings held considerable contradictions. For example, in a discussion of the student-produced list of "Rules for Graduate Students," faculty found it easy to understand this list simultaneously as a sign that students were immature and to express the idea that it was a good thing that students were cynical because academic life is hard. Indeed, as Barbara Lovitts's research on dissertation committee criteria across the disciplines repeatedly points out, there is a lack of consensus and considerable contradiction among faculty in how they develop standards of evaluation. This fits with the individualized, privatized model that the faculty were themselves trained in—they had no sense of collective or systemic agreement of the "rules" (Lovitts 2007).

Second, and in a related phenomenon, faculty and students had very different understandings of what happens over the course of a graduate school experience. Students tended to view becoming a scientist or mathematician as a constructed and somewhat arbitrary process. Faculty, in contrast, saw the process as natural, involving the growth and maturation of something already inside students. Faculty expressed an ideology of meritocracy, believing that there were those who are good at science and those who were not, those who had "it" and those who did not.

For example, the notion of a "spark" was an important part of faculty's description of their ideal students. Faculty largely believed that there was not a whole lot they could do to create this spark, but that once it was there, it was their role to mentor students toward successful scientific careers. Implicit in their narrative was the notion that the making of scientists was a natural process, involving the growth and maturation of something already inside students in incipient form. They felt this growth was idiosyncratic—something that sometimes happened for students and sometimes did not. Faculty felt they had only limited influence on it. Their understanding of what happens left little room for criticism of them as mentors, in the sense that it emphasized a "stay if you fit in, leave if you don't" perspective. That is, in faculty participants' accounts, a student should be able to tell that she or he is really "cut out to be a scientist" if the graduate education process comes to seem easy, reasonable, and rational. If it does not, the student was not meant to be a scientist. This framework (conveniently in the eye of students) short-circuited the possibility of critique. It also completely obscured the possibility of an inclusive pedagogy—an ethics that anyone can learn if taught!

Students in contrast were interested in challenging and reinterpreting the question of who could be a good scientist. They believed that good scientists could be made, and faculty should not rely on their perceptions on whether a student had a "spark." Students reacted strongly to the notion, suggesting that faculty were creating a mystical, nebulous quality to education ("like a mystic talking about attaining a plane of consciousness"), rather than addressing their role as educators in the making of scientists. Students interpreted faculty descriptions of mentoring as a disengaged process.

> It's very much a sink or swim attitude. As though they're just standing on the shore with their arms folded saying, "Yeah, that one is not going to make it. And that one is."

> If they told you that's what they were doing at least then while you're drowning you know. You could look up and see who was watching. Or you could say, "Oh! I am supposed to get out of this myself," but you know it's never made explicit and that's a problem.

THE "UNWRITTEN" RULES OF GRADUATE EDUCATION

What are the "written" rules of graduate education? There seemed to be surprisingly few. There were stipulations on required courses (with waivers if the committee agreed), deadlines on picking a dissertation advisor and topic, time for the comprehensive exam, the format of the exams, the dissertation committee and its composition, and limits on the length of time to the graduate degree (after which one had to retake comprehensive exams). But in each of these and in the rest of graduate student life, there was a surprising amount of flexibility. And indeed, there was tremendous variation in graduate student progress and performance. This sets up students to succeed in a very individualized model of achievement with a great deal of uncertainty around one's performance. I remember that when I was in graduate school, we were evaluated each year, and if nothing was amiss, we received a letter saying, "We have not found any evidence that you are not progressing satisfactorily toward your degree." Hardly a resounding endorsement!

One of the more powerful exercises we did was the rules exercise where we invited graduate student and faculty groups to name the unwritten rules governing graduate education. This exercise clearly encapsulated the fundamentally different standpoints that emerged among faculty and students. For this exercise we asked the four groups to name the unwritten rules of academia. Were there any rules that were not explicitly written down but widely believed to be rules or good practices by graduate students? If so, what were they? Each of the four groups was asked to generate the list without seeing the rules generated by the other groups. Anyone was allowed to suggest a rule. The rule was then discussed. If everyone agreed, it went on the list. Even if one person disagreed, the rule was left out. The wording was often carefully and painstakingly negotiated. The four lists of rules appear in the appendix to this chapter.

The rules exercise was much easier to do with the graduate students than the faculty groups. The students seemed to quickly grasp the point of the exercise, and there was easy agreement. There was more disagreement about the rules with the faculty group, and often this meant that rules were "worded down." Some of the faculty were uncomfortable with the strong wording and successfully argued to make them more general and open-ended. For example, "Do not have children" after a lengthy discussion was changed to "having children is an issue." Similarly, "pregnancy is a liability" was changed to "pregnancy and other time-consuming personal decisions are problematic." There was a striking difference in the level of consensus and the degree to which individuals felt there were clear rules. The differing tone and emotions in the room were palpable. If one faculty member suggested a rule, some other faculty member would counter it with an example or felt that was too rigid a stipulation. In the student group, by contrast, when a rule was suggested, they often felt like moments of revelation—that people had articulated something that had organized their lives for many years but no one had named until that point.

REACTIONS AND ANALYSIS

The faculty-student (via facilitators) conversations in our project highlighted the unique standpoints by which faculty and students experienced and interpreted scientific culture. Power implicitly shaped these differences in perspective just as power has demonstrably shaped the practice of science and the production of knowledge more generally. Ironically, and not surprisingly, initially neither group felt they had any power![3] One of the outcomes of the project was to get both groups to appreciate the power that they held, a power that could allow them to rethink and reshape the cultures they inhabited. As Foucault suggests, power is "exercised" in relation to others rather than a quality merely possessed (Foucault 1980). In our project, faculty and students experienced their power, or lack thereof, through relationships experienced through the graduate education process. The significance of the rules is not in whether these rules were empirically accurate or not. The significance lies in the implicit belief of the students in the rules, an implicitness absent among faculty narratives.

The rules exercise illustrates how gender norms and roles are operationalized in specific settings within graduate science and math education. The sets of discussions around the rules exercise provided a rich set of interpretations of the culture of science from participants. The culture of science, like all cultures, represents sets of rules and expectations of all individuals who participate in that culture. Social relations, and the rules and expectations that govern these relations in fields like science and engineering, have a long history of male domination and are particularly problematic for women. One of the strongest tenets in the culture of science is a studied silence about social relations, a silence designed to distinguish scientists from others. In our project, the resistance of faculty to talking about and recognizing larger patterns in their daily interactions and behaviors seemed a symptom of a more general resistance to their own cultural embeddedness. It suggested that those with power and privilege cultivate an exaggerated commitment to individuality as a form of resistance to cultural change (Subramaniam and Wyer 1998).

The expressions of privileged perspectives we found in our project were consistent with the concept of the "invisible knapsack" often used in introductory women's studies courses (McIntosh 2003). The nature of privilege, it is argued, is unmarked—an unweighted and invisible capacity to move without burden in a structure in which you are the norm, versus the burdens others carry of being marked. Scientific culture, students suggest, has unwritten rules and unacknowledged notions of who makes a good scientist. These characteristics of good scientists and science's "no culture" are deeply aligned through history with the west, masculinity, whiteness, and heterosexuality. It is precisely the privilege of *being* an individual that is critical, rather than a situation in which one's "individuality" is eclipsed by the markings of subaltern identity. One could argue that scientific claims to being culture free are not unlike a racist system's claims to color blindness—whereby the insistence that color doesn't matter is precisely what allows racial privilege to continue unchecked.

While many graduate women in our study were articulate about these institutional practices, it was apparent that initially they were reluctant to speak up, precisely because they felt surrounded by a culture of science that was silent on the importance of social categories of gender, race, class, and sexuality in the doing of science—in interpersonal dynamics, mentoring relations, and the role of personal issues in professional life. Students and faculty alike repeatedly remarked how much they learned purely by talking to one another about social relations, a practice they had seldom engaged within their home departments. The rules exercise provided a useful tool by which students and faculty examined the culture they worked in and in highlighting how similar and different those views can sometimes be. The exercise produced an interesting moment in knowledge generation as an actual *intervention* in the culture and practice of science.

In retrospect, the list of faculty rules more closely resembles the goals of graduate education than codes of conduct—they stressed the learning of the social and intellectual climate of one's field, learning to formulate and carry out research projects, and learning to communicate about one's work. As one student remarked, "Who could disagree with that?" Students interpreted their set of rules as the means and mechanisms by which they could attain the faculty's list of "goals." For example, faculty rules stress the following:

- Graduate education should be the major focus of your life.
- Personal lives and extracurricular activities are okay so long as they do not interfere with reasonable progress.
- Become the "expert" in your field.
- Show initiative and independence.

In order to attain the above goals, students pointed to rules in their lists that allowed them to meet the above expectations. These rules include:

- Work all the time with no break.
- Be visible around the department.
- Don't show fear, self- doubt, insecurity.
- Don't have a life outside the department (or don't talk about it).

- Be assertive, confident, good communicator.
- Be happy.
- Be busy.
- Don't complain even about real problems.
- Don't cry in public.

Thus students responded to the sets of unwritten expectations of faculty by a set of behaviors that best presented them as "serious" students. Students elaborated how gender was an important factor in their graduate school experiences and in their everyday lives. The rules exercise points to the perceptions women students have of their place in science. Women face an additional burden in having to follow the rules to be seen seriously as scientists. In their personal lives outside scientific culture, women face the pressure of conforming to our culture's notions of "femininity." But once they come to work, within scientific culture, they have to leave that identity behind, to assume another, namely, a form of "masculinity." Often, this takes the form of dissembling behaviors. Within the culture of science, women students participate in these rules in public, such as being happy or not crying in public. Students respond with their own set of rules in private, however—"cry in your office" or "leave your lab if you think you are going to explode." Student narratives unfold a world of performances, where they learn to perform a set of behaviors and practices that will allow them to be taken seriously. A "serious scientist" is single minded, dedicated, emotionless, happy, well connected, intellectually curious, the expert, and ready to follow in the footsteps of the mentor to reproduce the system. It is not that the students themselves change. The contradictions around emotions are always interesting—that emotions are always read as undesirable and feminine; hence it is possible to hold the contradiction of being emotionless and happy! Instead, they quickly learn to segment their lives—they cry in their offices or the bathroom, they learn to leave the lab to release their frustration or tension outside. And yet as the rules tell us, women scientists are not accepted completely if they become masculine. For no matter how much they give up femininity, they can never be men. Narratives that have emerged from the recent rise of the transgender movement, such as Ben Barres's, emphasize this point. This, I believe, is the double bind students describe and what Keller argues when she talks about women scientists being bounded by inauthenticity and subversion (E. Keller 1987).

The faculty were of course deeply disturbed by these rules. Their responses were threefold. One was complete refusal to believe that these rules exist. Some specifically talked about how they work against these rules especially by being models themselves. It remained a consistent struggle in discussions to have faculty elaborate a role of a collective culture even though they may have individually resisted specific rules. The second response was that students were missing the point. They thought that if students did not enjoy doing science, did not enjoy participating in these rules, that rather than dissembling or pretending or play acting, they ought not to pursue this line of work. A third response (especially around the persistence of the long work hours, the importance of connections, and the chilly climate for women) was that students had gotten it right and life would not get any easier when students joined the faculty ranks! Again, while there were differences between men and women faculty, especially when many women faculty felt marked by their own gender, there were far more similarities between men and women faculty than between women faculty and students.

With respect to gender, it seemed that students had a more unitary and cogent understanding of how gender impacted their lives as students and scientists than faculty did. Yet women faculty were quite explicit and active in enforcing gender norms, particularly with respect to dress (not too feminine, not too casual). Even so, it seemed that departments did evolve their own idiosyncratic cultures, and overall, gender norms and expectations were more coherent within a single department than across departments or the university as a whole. This, in turn, provided a clue as to why student and faculty interpretations might have differed. For students, power tended to be concentrated in the single form of the faculty mentor. For faculty, in contrast, power was much more diffused—in department heads, senior colleagues, administration, journal editors, professional colleagues at conferences, and so forth. Thus, faculty indeed experienced gendered pressures and inequalities in more diffuse contexts. Even pretenure faculty, who are perhaps the most vulnerable, recognize the need to impress all their faculty colleagues in contrast to students who are primarily accountable to their individual dissertation advisors.

It seemed to us, in listening to the two groups describe graduate education and the day-to-day life of graduate student-faculty mentor interaction, that both significantly misapprehended their relation to the other. Graduate students tended to overestimate faculty power, control, and investment in the day-to-day life of graduate students. They also reported that it was hard for them to understand their place in the wider academic system. They had difficulty understanding how graduate education fit within the larger set of constraints and obligations experienced by their faculty mentors. Faculty, in turn, tended to underestimate their power in relation to graduate students, their impact on the quality of life of their students, and their role as gatekeepers to professional careers in the science and mathematics.

As feminist scholarship reminds us, gender is something we learn to "do," not an attribute one "has." For the students in our project, graduate education was a contested exercise in enculturating students into scientific culture and teaching students how to "be" scientists. What women students in the project were often challenging and objecting to in their descriptions of the rules of scientific culture was not an unwillingness to follow the rules to do science. Rather, they challenged the notion that these rules did produce good science and, more important, that *not* following these rules would produce bad science. They weren't questioning the need for structures themselves, but critiquing the values attributed to those rules and the ways they dispersed privilege unevenly. Empirical work on the qualities of good scientists would suggest that there is no ideal constellation of cognitive abilities that shape the ideal scientist. Successful scientists demonstrate a range of characteristics such as deductive reasoning abilities, verbal skills, quantitative reasoning, intuition, and social skills (Handelsman et al. 2005). Narratives also highlight creativity, imagination, effective communication, and thinking outside the box as useful skills (Handelsman et al. 2005).

If the current rules did not necessarily produce "good" science, then we were free to imagine alternate frameworks. This re-envisioning allows us to draw on models that are diverse, flexible, and informed by difference. It was exciting to see that by the end of our project, participants were imagining alternate frameworks for graduate education in the sciences. By examining and critiquing their own culture, by the end of the project, faculty and students began imagining and exploring and ultimately even implementing alternate models for a culture of science, one that did not define science and scientists exclusively in terms that our culture has relegated to the "masculine."

Such an understanding of the culture of science can be immensely powerful for the transformation of institutions as well as those of individual students and faculty. Rather than train graduate women to perform successfully in the existing culture of science as most mentoring programs do, we can instead asked men and women in science, faculty and students, to examine the world in which they live. Through the articulation of what we do and how we do it, we can ask ourselves how we can do it differently—incorporating and celebrating the diversity of genders, ethnicities, classes, nationalities, and sexualities that make up our world. Perhaps, in so doing, one day the diversity of scientists will mirror the composition of that diverse world, and in so doing, produce very different—fuller, multifaceted, interdisciplinary science.

Appendix. NSF Model Project

FACULTY GROUP I

Some Unwritten Rules of Graduate School/Academia

1. Think of yourself as an independent researcher as early as possible.
2. Show initiative and independence.
3. Graduate education should become the major focus of your life.
4. You should decide early if you enjoy research and if not find something else.
5. Having children is an issue, although it's getting better.
6. Pedigree is still important.
7. Read more, think more than you are required to.
8. Go to seminars.
9a. Develop communication skills—give talks/seminars.
9b. Learn to write.

FACULTY GROUP II

Unwritten Rules of Graduate School

1. Attend seminars/colloquia.
2. Participate actively in seminars (at least eventually).
3. Be intellectually curious—take initiative, ask questions, visibly discuss science with faculty/students.
4. Take responsibility, be professional (you are not an undergraduate anymore)—work hard, make progress.
5. Take ownership of your research project—become an "expert" in your field.
   - Learn the background.
   - Be independent.
   - Take intellectual responsibility.
   - Learn how to locate information.
6. At least begin to learn the social and intellectual climates of your field (including outside the home institution).
7. Research/academic positions are construed as "best."
8. Exhibit integrated, clear thinking.
9. Cooperation and collegiality (exhibit it).
10. Don't procrastinate too long—realize that graduate school is a life phase, not a lifestyle.
11. Practice and develop communication skills—oral and written.
12. Learn to communicate about your own work.
13. Personal lives and extracurricular activities are okay as long as they do not interfere with reasonable progress.

14. Pregnancy and time-consuming personal decisions are problematic.
15. Personal and professional needs have to be balanced in a constructive way.
16. Don't expect your research advisor to be your personal and professional mentor.

STUDENT GROUP I

Unwritten Rules of Graduate School

1. Work all the time with no breaks.
2. The only work that counts is what is publishable.
3. Students on the fast track will get more resources and faculty time.
4. Emotions permitted; excitement, "controlled" emotions not allowed: being upset, overwhelmed, insecurity, apathy, and lack of interest.
5. Don't go public with career plans outside Research 1 university jobs.
6. Be like your advisor.
7. Connections count (good and bad).
8. Don't show fear, self-doubt, insecurity.
9. Don't have a life outside the department (or don't talk about it).
10. Be assertive, confident, good communicator.
11. Know your place in the hierarchy.
12. Students ought to be available whenever faculty want, faculty will be available to students whenever faculty want.
13. Faculty time is more important than student time.
14. Students should be more visible around the department (not true in Math).
15. Know the local landscape about rules.
16. If things aren't going well, you may not get direct feedback.

Rules for Women Graduate Students

1. If you get pregnant, you may be perceived as less serious but if you surmount that, kid(s) must never interfere with work.
2. Double binds: know your place and don't be timid.
3. You will be either a bitch or a doormat (more or less true depending on how many women are around).
4. Not okay emotions—being upset, feeling defeated, overwhelmed, insecurity, apathy, self-doubt (more true for women than men).

How do you get around the rules?

1. Ask permission.
2. Ignore the rules (can't always get busted because they are unwritten).

198

3. Pick and choose rules you want to follow (get help from other graduate students).

STUDENT GROUP II

Unwritten Rules for Graduate School

1. The official rules are flexible.
2. Early success helps.
3. Be visible at departmental functions—be a "good" citizen.
4. Be a colleague to faculty, but know your place.
5. Don't complain, even about real problems.
6. Don't make waves—don't be a troublemaker—you will be punished (this extends to junior–low-status faculty).
7. You don't have input, even on decisions that affect graduate school (even when asked).
8. Faculty will not interfere in other faculty's business.
9. Connections count.
10. Be happy.
11. Be busy.
12. Be visible in the department.
13. Don't cry in public.
14. Work all the time.
15. Be just like your advisor.
16. There is an "expected" track, but the funding crunch is modifying it.

Rules for women graduate students

1. Act happy.
2. Cry in your office.
3. Don't complain (especially true for women).
4. Dressing "feminine" elicits less respect, more help; your legs are public property.
5. Being a strong woman, a strong advocate for women, or notably a feminist is not good either.
6. Being a woman is a liability. You have less leeway with unwritten rules if you are a woman (varies with department).
7. Don't be feminine: don't exhibit—dress, intonation of voice (feminine speech), don't wear makeup, don't show feminine emotions (i.e., women more vocal about emotions, menstruation).
8. Pretend to be like your advisor or be quiet about it.
9. Pregnancy is a liability (pulls you off the expected career track).
10. Don't have a personal life.

CHAPTER NINE

The Emperor's New Clothes

Revisiting the Question of Women in *the Sciences*

"The boy is right! The Emperor is naked! It's true!" The
Emperor realized that the people were right but could not
admit to that. He thought it better to continue the procession
under the illusion that anyone who couldn't see his clothes
was either stupid or incompetent. And he stood stiffly on his
carriage, while behind him a page held his imaginary mantle.

—Hans Christian Anderson, *The Emperor's New Clothes*

The Lesson

And after the Emperor had appeared naked and no one had
disturbed the solemn occasion, one little girl went home in
silence, and took off her clothes. Then she said to her mother,
"Look at me, please, I am an Emperor." To which her mother
replied, "Don't be silly, darling. Only little boys grow up to be
Emperors. As for little girls, they marry Emperors; and they
learn to hold their tongues, particularly on the subject of the
Emperor's clothes."

—Suniti Namjoshi, *Feminist Fables*

In the famous Hans Christian Anderson fable, *The Emperor's New Clothes*, the
emperor commissions a cloak. Under pressure to produce ever finer cloaks,
the cloth maker and tailor invents a deception—the finest fabric ever, a fabric
so fine and fantastical that it is visible only to the truly intellectually deserving
and is invisible to all else. While the emperor himself cannot see the cloak, he
pretends to be able to and wears his imaginary cloak in a procession through

town. The "emperor's new clothes" has become a metaphor for power, pretentiousness, social hypocrisy, and collective denial. I use the emperor's cloak as a useful metaphor to discuss "cloaks" that frame the literature on women in the sciences. As the famous fable unfolds, the cloak turns out to be nothing but an illusion—nonexistent and fantastical. It is only given visibility through the power of the emperor and the complicity of his subjects; the emperor is, in fact, quite naked.

Having examined variation's genealogy in evolutionary biology and traced its ecological geographies, in part III of this book I have been looking at how the question of variation has haunted our ideas of who can be a scientist and has shaped demographic patterns of scientific practitioners. Ideological debates about the importance of variation for evolutionary theory seeped into social relations within science in unexamined and disquieting ways. Women, as "outsiders" who might have provided variation in perspectives within science, have been in some historical periods expressly excluded, in others belittled and discounted, and in more recent times silenced and implicitly excluded through the micro-dynamics of "cumulative disadvantage" (Etzkowitz et al. 1994).

Nonetheless, women and other marginalized groups have fought long and hard to enter the hallways of science. In the past three decades activists have moved through the courts with antidiscrimination laws and affirmative action programs (such as Title IX in the United States) as well as through women in science and engineering (WISE) initiatives, projects, and programs. By WISE initiatives, I include programs literally named WISE at most universities but also women in science initiatives at professional societies and institutions, that is, funded programs and projects on "women in science and engineering." WISE initiatives have their political and intellectual roots in the women's movement and its academic arm, women's studies. Indeed, we can trace the origins of what is now called feminist science studies to early feminist work in the late 1970s and early 1980s when women scientists began to articulate how questions of sex, gender, race, sexuality, and nationality were inextricably interconnected and implicated in the institution of science, its history, cultures, people, practices, and knowledges. Despite these origins in the experiences of women scientists, feminist work within women's studies has gradually moved away from questions of the "problem" of women *in* the sciences to ask questions about gender *and* science. The literature on gender and science focused not on the practitioners of science but on the ways in which scientific knowledge was gendered; work shifted to the important project of exploring the production of knowledge in science. The move from women *in* the sciences to gender *and* science divorced questions of the practitioners and producers of scientific knowledge from questions about the scientific knowledge they produced. Scholars focused on the content of science, while activist faculty and staff focused on providing a supportive environment for students who were majoring in science, technology, engineering, and math (STEM) fields. The activist efforts resulted in the growth of WISE programs, largely housed within the administrative wings of science and engineering colleges. These programs have been broadly charged to increase the recruitment, retention, and overall status of women in the sciences. While their charge is broad, the focus is usually more instrumental than academic, in providing women with networking and mentoring opportunities, professional development activities, and leadership training as part of an equity feminist approach. They seldom articulate a critique of androcentric bias in academic practices, and instead offer a "you can do it" cheerleader approach to women students touched by the program. The programs are also usually evaluated by the popularity of events rather than the institutional change they have effected.

I am myself deeply indebted to WISE programs. They were an important part of my early beginnings and I have since worked in and continue to support WISE programs. Yet, as a woman in science who works in feminist science studies, I am struck by how little interaction there is between the two. This chapter is written in the spirit of the tremendous possibilities of coalitions and collaboration that I see between women's studies and WISE initiatives. At this time, with few exceptions, WISE programs tend not to be centrally located in feminist science studies or women's studies programs and departments since they are not engaged with questions of scholarship or curriculum. Given their broad mission, WISE programs could have developed a robust analysis. As I elaborate later in this chapter, it should not surprise us that they have not. Women and gender studies programs, in turn, have too few allies in science and engineering fields. While we could debate the details of institutional dynamics, one thing is clear. While all three fields—WISE programs, feminist science studies, and women/gender/sexuality studies—can arguably be said to be thriving, there is little interaction between them. The considerable funds invested in WISE programs and initiatives could, I believe, yield impressive results if WISE programs engaged more fully with feminist work in the social sciences and the humanities. If we are truly interested in diversity and addressing the question of variation on all levels—genealogical, geographic, and biographic—it must involve bringing the project of women *in* science back together with the question of gender *and* science.

Between the Matildas and Curies:
A Brief History of Women *in* the Sciences

In challenging the claims of scientific objectivity, feminists have uncovered the systematic ways in which power and privilege have shaped science and scientific practice. They have detailed the systemic exclusion and discrimination against women.[1] These exclusions were based on a complex stew of ideologies about rationality, androcentrism, white privilege, eurocentrism, and biology. It is now apparent that white males dominate the scientific elite through no accident of history, as scientists were powerful cultural and institutional voices in sustaining assumptions about white male superiority in science—hardly disinterested or objective. In this section, I briefly review the history of women in the sciences in the United States—the dynamics of exclusion, as well as the social movements by women scientists to challenge these exclusions and demand their right to full participation in the scientific enterprise. There are two predominant concepts in this history of women in the sciences: the "Matilda effect" and the "Curie effect." The Matilda effect was named by Margaret Rossiter as a corollary to the Matthew effect (Rossiter 1993). The Matthew effect contends that privilege matters within science and that the work of highly ranked scientists is more likely to be noticed and recognized than that of lesser-ranked scientists (Merton 1968). Rossiter coined a term for the inverse effect, the *Matilda effect,* which argues that in contrast to the Matthew effect, the contributions of women scientists are usually denied or ignored. After their deaths, they are forgotten, their contributions lost and invisible in the historical records of science. The effect is named after Matilda Gage, an important American suffragist who argued that the inventor of the cotton gin was a woman (Hess 1997). In contrast to the Matilda effect, we have its converse, the Curie effect, which recognizes the popular conception of Marie Curie as *the* woman scientist (Des Jardins 2010). The Curie effect points out that in order to be famous, women scientists have to be unusually brilliant like Marie Curie, who won a Nobel Prize in the sciences not once but twice. These two effects represent two phenomena in the history of women in the sciences—invisibility and extraordinariness. Somewhere between the forgotten Matildas and the extraordinary Curies lies the worlds of most women in the sciences.

Nonetheless, historians have unearthed the long buried contributions of women to science, mostly focused on the United States and Western Europe (Gates and Shteir 1997, Shteir 1999). This is not a linear history where women have been accepted into the profession with increasing and growing respect.

Rather, political shifts have influenced science and women's contributions in complex ways (Rossiter 1982, 1995, Eisenhart and Finkel 1998). Historians have unearthed biographical details of "women of distinction" (Kohlstedt 2004: 3) such as Rachel Carson (Hynes 1989, Lear 1997), Anna Botsford Comstock (Henson 1987, Kohlstedt 2004), Marie Curie (Pycior, Slack, and Abir-Am 1996, Pycior 1997, Jardins 2010, Quinn 1995), Rosalind Franklin (Sayre 1975, Maddox 2002), Anna Mani (Sur 2001, 2011), Mileva Maric (Renn and Schulmann 2000), Barbara McClintock (E. Keller 1983), Lise Meitner (Sime 1997), Anna Kouzeletsova (Koblitz 1983), and Ellen Swallow Richards (Lippincott 2003), among others. Questions of nation, race, and class complicate these narratives (Koblitz 1983, Sur 2001, Johnson 2011, Bilimoria and Liang 2011).

Margaret Rossiter's now classic two volumes on women in science as well the work of other historians demonstrate that the systematic marginalization of women has kept the numbers of "great women" to relatively small numbers, that is, exceptions rather than the rule (Rossiter 1982, 1993, Abir-Am and Outram 1987, H. Rose 1994, Pycior et al. 1996, Henrion 1997). A wide variety of studies have revealed a consistent pattern of discrimination. Several influential phrases frame these perspectives (Wylie 2011)—women in academe face a "chilly climate" (Hall and Sandler 1982, 1984, Sandler 1986), women find themselves in the "outer circle" (Zuckerman and Cole 1991), or women are "outsiders in the sacred grove" of the prestigious academy (Aisenberg and Harrington 1988); if they do succeed, they quickly hit the "science glass ceiling" (Rosser 2004a). Programs aimed at women in the sciences are predicated on the idea that women will make excellent scientists if only given the opportunity.

The national statistics showing an increase in Ph.D.s awarded to women in most STEM fields provide concrete evidence that women have been determined to become scientists, despite obstacles. The success of women in the sciences in the United States owes a great deal to their persistent organization and activism. Kohlstedt (2004) notes key successful strategies developed in the late eighteenth and early nineteenth centuries, which continue to this date: (1) documenting discrimination against women scientists (Rossiter 1982, S. Levine 1995, Kohlstedt 2004); (2) identifying practices and standards used to exclude women (Rossiter 1982, 1993, Kohlstedt 2004); (3) determining practices used to contain women in socially appropriate areas of work such as home economics or some fields in the biological sciences (Nerad 1999, Kohlstedt 2004, Veit 2011); and (4) developing remedies to redress the marginalization through public recognition of women, collective action, and leadership development (Kohlstedt 2004). While these strategies facilitated women receiving advanced degrees, they often failed to get jobs, promotion, and recognition. By the 1920s, women's advancement seemed to plateau or even lose ground (Kohlstedt 2004). The growing professionalization of scientific fields developed cultures of growing masculinity that further marginalized women (Nye 1997). As old barriers fell, new barriers and forms of stratifications were put in place (Rossiter 1982). During and after World War II, women in the sciences continued to be paid less than men and had restricted and even reduced access to laboratories and resources compared to wartime levels. The push to return women back to the home in the postwar years further exacerbated these patterns as mainstream ethos highlighted the role of women inside the home as wives and mothers (Kohlstedt 2004, Rossiter 1982).

The consistent activism and strategies had a cumulative effect, however. The postwar baby boom led to an increase in the number of women scientists although most were employed as researchers and faculty in the poorest and least prestigious institutions. Those in more prestigious universities were clustered in home economics, though a select few had appointments in genetics and anatomy (Kohlstedt 2004, Rossiter 1995).

The post-*Sputnik* support for science along with the rise of the women's movement and the emergence of the academic field of women's studies facilitated the organization of women's groups in science.[2] Social transformations in the workplace through the Equal Pay Act, push for childcare, and legal challenges to discrimination were important for all women, including scientists (Freedman 2002). Scholars began to document how precipitously women's presence in science declined with increasing educational achievement and used these statistics to push for programs at NSF and other institutions (Rossiter 1995). Equally striking was the unevenness in the presence of women— clustered in particular fields like psychology and the life sciences but severely underrepresented in others like the physical sciences and engineering. The metaphor of the "pipeline" emerges in this era, a metaphor about the participation of women in the sciences that has endured in the United States ever since.

By the 1990s, feminists gathered data through various federal agencies to show that the gains of previous eras were not being sustained (Kohlstedt 2004). Scientists and activists pushed for legislative and institutional commitments to change. Congress passed the Women in Science and Technology Equal Opportunity Act in 1980 extending Title IX law (which forbids sex discrimination at educational institutions that receive federal funding) to the fields of science and engineering (Handelsman et al. 2005). Regular surveys of the scientific workforce and yearly compiling of data on undergraduate and graduate degrees (organized by gender, race, nationality, and ability) by NSF gave rise to constant quantitative and qualitative monitoring of women and minorities in science (Mervis 2000).

Alongside promoters of women and minorities in science have been critics of such programs. They have argued that women are not good at science; that women don't want to do science; that such initiatives shortchange boys; or that affirmative action initiatives are inherently unfair (Cole 1979, Lawrence 2006, Pinker 2002, Baron-Cohen 2003, Mansfield 2006). These debates have often pitted men against women, whites against students of color, or U.S. nationals against foreign nationals. Questions of parity and fairness have been presented as being at odds with each other, and it is claimed that diversity is being promoted at the cost of excellence (Mervis 2000). These debates have often created a backlash against initiatives to increase the representation of women and other marginalized groups (Holden 2000). The 1990s also revealed a plateau in the increasing number of women in science, while the number of men declined. Women remained concentrated in the psychological and behavioral sciences (Kohlstedt 2004), and continued to be clustered in less prestigious institutions, outside the ranks of tenure and in institutions that focused on teaching rather than research (Long 2001). In fact, in the 1990s women lost ground in mathematics and computer science, fields that emerged as the "hot fields" of that decade (Kohlstedt 2004, National Science Foundation 2002). The 1999 Massachusetts Institute of Technology (MIT) report "Women in the School of Science" reframed the public debate about the status of women in science by powerfully demonstrating the continued discrimination in the post–civil rights era (Wylie 2011, MIT 1999). Discrimination, the report argued, was created by a pattern of "exclusion and invisibility" affecting women's workplaces, their quality of work life, and their careers and career trajectories compared to similarly well-trained male scientists (MIT 1999: 8).

While the above brief history has focused on the United States, underrepresentation of women in the sciences persists throughout the world, although the historical patterns vary considerably across continents and national boundaries. Women are underrepresented in the sciences across the globe, especially in the more prestigious and well-paying fields. In national contexts where mathematics and physics appear to be less important, women are better represented and more visible. Cross-national analyses make a strong case against biologically determinist arguments and demonstrate powerfully the complexities of the demography of the scientific workforce (Mellström 2009, Kumar 2009, Subramanyam 1998).

Thus, women's marginalization in science is best understood as one of political and systemic exclusion. These practices of exclusion have been well documented by women scientists and their biographers, historians and social scientists, lawyers and university administrators, and feminist science studies scholars. The practices and processes that have discouraged, marginalized, and diverted women from their scientific interests and talents are alarmingly enduring, evident in the research that spans over two centuries of U.S. history. The ideological thematic that underwrites this durability is the recurring debate about women's intellectual and physical difference from men, a difference constructed against a western-white-male-as-norm backdrop. The touchstone points in this debate echo and reassert twentieth-century eugenic arguments about the importance of variability in evolution. Of most concern are the ways in which today's discussions about women's abilities are voiced by those who intend to be advocates for women in science, even while their arguments evade confronting the deep structural foundations of masculinist science. It is an unavoidable fact that elite white western male scientists embedded notions of "superior" and "inferior" human biology into early formulations of evolution. Contemporary versions of these ideas are evident in resorting to discussion of demands of pregnancy, motherhood, and family as particular burdens that women in science do (or do not) or should (or should not) embrace. The case for biological difference endures in these formulations. Do women have the intellectual ability to be scientists? Can women's brains grasp the advanced mathematical modeling necessary in science? Can women have careers in science and be mothers too? Discussing and answering these questions is much-tread territory within programs for women in the sciences. Unfortunately, such emphases, in the absence of critiques of cultural norms in science, sustain the unspoken assumptions that variability is of questionable value for humanity.

Women in Science and Engineering Programs and the Politics of Inclusion

Considering that women were once seen as incapable of rational thought and were barred from higher education and the hallowed halls of science, the success of women in STEM fields today is remarkable. The proportion of women in science and engineering occupations has increased considerably (Bilimoria and Liang 2011). For example, there has been a thirty-fold increase in the proportion of Ph.D.s granted to women in engineering. This period corresponds with women's persistent activism; this shift is best explained by the removal of cultural and structural impediments rather than innate differences (Handelsman et al. 2005). Indeed, we should not underestimate these changes. Yet, as we shall see, the increase has plateaued in recent years or at times, even declined. Why?

Over the past three decades, government agencies such as NSF, the National Institutes of Health (NIH), as well as private organizations such as the Association of Women in Science (AWIS), the American Association of University Women (AAUW), the Women in Engineering Professional Action Network (WEPAN), the Society of Women Engineers (SWE), as well as women's task-forces and caucuses within professional societies of scientific disciplines have all funded projects and programs to increase the recruitment and retention of girls and women in the sciences. There is an enormous literature that explores the successes and challenges of various intervention efforts at K–12, undergraduate, graduate, postgraduate, and faculty levels.[3] The pattern of underrepresentation of women persists for women in most science settings in contemporary times (National Academy of Sciences 2006, Settles et al. 2006), and studies repeatedly find that bias persists (Kenneth Chang 2012, Moss-Racusin et al. 2012). Targeted programs at NSF have worked to intervene in the institutional culture of departments (Rosser 2004b): working to reduce discriminatory practices in hiring and promotion (Fuller and Meiners 2005, Martinez et al. 2007, Rosser 2004a), as well as to empower and mentor women for a successful life in science and engineering.

To the often-asked question on whether feminism has changed science, most have answered in the affirmative but with qualifications (Bug 2003, Kass-Simon and Farnes 1990, Rosser 1990, Schiebinger 1999, 2008). Women's presence has not changed science because they have brought feminine or feminist values but because their very presence has helped erode traditional gender labels (E. Keller 2001). Indeed, despite well-developed theories and analyses on feminism and science, initiatives for women in the sciences have been grounded in and reduced to a narrow liberal idea of equity or equality of representation as the primary intellectual, political, and ideological strategy (Cinda-Sue Davis 1996). This comes from a larger cultural adoption of a principle of equality of *opportunity* rather than equality of *outcome*—which is the more radical commitment to social justice. It is well worth pondering why demographic numbers are the cornerstone of focus rather than cultural or structural transformation. Despite equity feminism that most of these efforts promote, parity for women in the sciences remains a distant goal—the numbers seem to have plateaued, and in some fields and subfields they have declined (National Science Foundation 2011). The professoriate remains male dominated, especially at prestigious institutions and senior ranks. Why? At the heart of the problem here is the very worldview that frames women in science discourse. The problem, as the pioneers of feminist science studies articulated it, is an exclusive focus on women instead of on the gendered and racialized nature of the institution of science itself. The exclusive focus on increasing numbers of women has led to a myopic and singular focus on equity and parity and a narrow vision of feminism and what feminist theories can offer. The women *in* science discourse has been haunted by a path of "relentless linear progressivism" as the result of particular discursive formations (Garforth and Kerr 2009). There have been some changes in the predominant language in the past three decades, but interrogating the underlying assumptions about women is rare. The shift from exclusion to inclusion is marked by a shift from exclusion based on claims of the innate biological inferiority of women's scientific abilities to a politics of inclusion dominated by policies that address women's biological bodies and gendered roles as wives, daughters, and mothers. Consistent with other segments of the labor market, in considering the various barriers facing women, women scientists most often cite the need to balance career and family (Rosser 2004b). Interviews, case studies, and statistical research consistently find that individuals report that family/work balance discriminates against women scientists at structural, institutional, and individual levels (Rosser 2004b, Rosser and Taylor 2009). The personal choices, relationships, and responsibilities of women outside the halls of science (especially as wives, mothers, and daughters) have nurtured and supported women's scientific work but also stymied and curtailed careers (Kohlstedt 2004, Zuckerman and Cole 1991, Laslett and Thorne 1997). Pregnancies, childcare, and housework have always largely fallen within the domain of women's work and women's roles as wives and mothers, and so have been consistently highlighted as a reason for women's lack of equal participation (Long 2001, Mason and Goulden 2004, Xie and Shauman 2003).

These concerns have led to a push in "female-friendly" policies. Despite their progressive ambitions, emphasizing issues of reproduction and family, advocates of "family-friendly" policies reassert women's reproductive potential as a central concern, marking "female difference" as hypervisible while leaving the worlds of masculine epistemic cultures untouched. To be sure, it is neither desirable nor persuasive to articulate an "anti-family-friendly" perspective at this moment in time, but as a strategy for inclusion, the consistent emphasis on family and women reinforces essentialist ideas about women. What has remained unchallenged is the normative model of the male as *the* ideal scientist, which insists on a productivity that can only be achieved by very long hours, a singular dedication to work, and an exclusive focus on one's profession. Solutions have included mentoring women to negotiate the normative model, promoting those who accept the normative model, retaining women through "special accommodations" that increase their workplace flexibility (part-time appointments with administrative or teaching responsibilities), automatic pregnancy leave, and family leave. The original standards for excellence are never challenged; the solution is about helping women conform to them. In extolling feminine virtues, the women in science discourse endlessly reinscribes women in relation to femininity and the domestic reproductive sphere (Garforth and Kerr 2009). The idea that the inclusion of women means first and foremost "family friendly" creates a universal woman and positions all women in relation to a heteronormative reproductive economic model (L. Morley 1999, 2006). In addition to essentializing women as different from men and without variability from one another in commitments to family and children, such strategies leave the mythology of an objective, value-free masculine world of science and technology untouched (Garforth and Kerr 2009). Indeed, as Louise Morley (2003) has shown, at higher levels of science, such a focus on women and family has created a hypervisibility around women, gender, and family, rendering women more vulnerable in their careers as their success is often seen as political rather than meritorious. In all of this, women get marked as separate and their bodies are marked as different and ones that need to be endlessly monitored, while men remain invisible and unmarked. After all, as feminists have long argued, gender is not what one "has" but what one "does" (Morland 2011). Challenging the elision of norms of masculinity and scientific masculinity has to be a central project for feminism and science.

Finally, initiatives related to women in science and engineering (WISE) have created an endless structure of vigilance and monitoring (Garforth and Kerr 2009). There are constant audits, grounded in identity categories of gender and race, and these evaluations constantly reinforce and reinscribe rather than challenge a rather narrow vision of race and gender. Again and again, gender is discussed in relation to work/life balance, childcare, harassment, bullying, human resources, recruitment, promotion, and so on. At the same time large swathes of epistemic, organizational, and personal lives of institutions and researchers remain unquestioned, including normative expectations, research policies, male networks, and privilege (Garforth and Kerr 2009). When inequities persist and the initiatives fail, they are seen as "implementation gaps." Affirmative action and equal opportunity policies become audits and an "inspecting body" rather than a challenge to the micropolitics of institutional power. These efforts have ended up making the identity of marginalized individuals more visible and their presence endlessly surveilled. Through all of this, masculinity and whiteness remain the unmarked and normative categories that escape scrutiny. As Garforth and Kerr (2009: 398) eloquently summarize: "Perhaps the problem with women in science is not women or science, but a women in science problem in its own right."

210

In examining the history of initiatives for women in the sciences, I consider *five* key problems that haunt efforts on behalf of women in the sciences.

METAPHORS THAT FRAME THE DISCOURSE ON WOMEN IN SCIENCE

One of the key insights of feminist science studies is that language matters (E. Keller 1985). Language embeds, expresses, and enacts the concepts that frame an issue, thereby shaping resulting policies and solutions. The WISE literature is rich with governing metaphors. Catch phrases and pithy summaries reveal a great deal about the ways in which interventions have been framed and characterized. Two related metaphors govern the field—the process of producing talented scientists as a pipeline that "leaks" and the process of career development as a "path" that leads necessarily to success albeit with some bumps along the way.

The "pipeline" metaphor creates a visual flow-chart about those entering and leaving the sciences and is widely used in discussions of the recruitment and retention of girls and women in science (Subramaniam 2009). The metaphor invokes a long pipe leading from kindergarten to the scientific laboratory with "leaks" representing the attrition of women and students of color. Solutions are conceptualized as "plugs" that keep more women and students of color within the institutional machinery of science. The pipeline model is predicated on the notion that if more women enter the pipeline and stay in it, more will become scientists and academics. Low participation in science and attrition from science are portrayed as resulting from individual choices rather than a process of discrimination. Such a portrayal fails to account for structures of institutions or practices of science (Schiebinger 1999, Bystydzienski and Bird 2006) or social influences and contexts (Xie and Shauman 2003).

While the metaphor of the pipeline describes flows in the machinery of science, it also evokes other visions. We could describe the pipes as long, dark, dingy, impenetrable tubes and masses of metal crisscrossing the terrain of industrial capital; pipes contain, constrain, limit, and cut off the oxygen from the travelers within. Imagining the regimented travels in pipes that give the travelers no agency in their journey, we might start cheering for the leaks and for those who escape the drudgery of pipe travel! Such a picture also describes the experiences of many women in the sciences, although, I hasten to add, not their thrill of discovery and exploration (Subramaniam 2009). And this, I believe, has been the crux of the difference between the literatures on women *in* versus women/gender *and* sciences. In the former the leaks are seen as a problem, and in the latter the problem is the pipe itself. If science is indeed in the business of laying down pipes that are inhospitable to marginal groups, why encourage young girls and women to enter them or stay in them? Why not rejoice at the leaks as a symbol of escape?

The second metaphor represents a scientific career as the final destination along a clear path of steps, aka "career ladder." Here, I draw on the work of Garforth and Kerr (2009), whose incisive and insightful critique of such programs in the United Kingdom translates well to the United States. In their analyses, career development in science is presented as a linear and hierarchical progression in the growth of expertise, experience, and vision, all in a timely fashion. There are clear "stages" that are most often imagined as a "ladder" or "path" along which scientists need to move toward excellence. Resting, stopping, or not progressing "on time" is evidence of failure (Garforth and Kerr 2009). The focus is thus on counting bodies at every stage of the career path—K–12, undergraduate, graduate, postdoctorate, assistant professor, associate professor, professor, senior membership in elite professional societies, and recipient of awards and accolades. There are also references to challenges confronting marginalized groups along the way, describing an obstacle course—bumps, hurdles, hoops, and pitfalls as well as institutional obstacles such as glass ceilings.

As Garfield and Kerr (2009) argue, the WISE literature is organized around two related narratives—one about progressive change over time and another about science as a meritocracy. In the progressive change narrative, women in the sciences march toward equality within institutions that are becoming more modern with equal opportunity and greater diversity. In this narrative, historical change is "progress," where "much has already been done," inspiring advocates for women in the science to continue the work because "more needs to be done for full equity." In this narrative, a "more equitable future is always in sight" (Garfield and Kerr 2009: 389). In the second narrative about science as a meritocracy, scientific institutions and their policies are seen as well structured and fair. It is individuals within them who are sometimes discriminatory. Through adequate training and mentoring, women can be taught to be more productive and assertive and the evaluators more objective and fair. Programs are largely focused on identifying women and people of color who "pass" or qualify for normative standards for insider status (Iverson 2007). Thus, it is assumed, that individuals with talent will inevitably (with hard work and dedication) be recognized and rewarded for their career achievements.

Both narratives eerily echo themes in the debates about variation in nineteenth-century evolutionary biology that asserted evolution as a historical progression toward excellence in a world filled with natural (and therefore desirable) hierarchies, with white males perching at the top. Evolutionary biology has yet to resolve the core debate between those who argue that variability promotes vitality and those who argue that variability dilutes it. The narrative of women *in* the sciences replicates science's own narrative of scientific progress, when in fact rather than linear progress, the history reveals circularity because the core problem of variation and difference is never solved. Thus the analogies of evolutionary biology's genealogy of variation mirror its biographic problems. Change-oriented activists on behalf of women in science would benefit from engaging with feminists who study the social world of science if they are to confront the narrative frameworks that contain their initiatives. For despite good intentions, strategies to include women have been governed by uninspired, regimented, and conformist notions about the conditions that foster a career in science. The programs work hard to enculturate women to survive the dingy recesses of the pipeline armed with strategies of self-motivation, networking, mentoring support, and superficial antidiscrimination policies to soldier on in a prescribed, timely fashion. Programs are inevitably focused on supporting women and people of color; the onus is always on women and people of color, not the structural forces of sexism, or racism, or other forms of discrimination (Iverson 2007). The western-white-male-heterosexual normative standards of science remain uninterrogated. If science developed as a world without women, women in science programs continue this by defining the goals of women's "success" in terms modeled on male scientific careers and female reproductive capacity.

THE CRISIS IN HIGHER EDUCATION AND THE QUESTION OF WOMEN IN SCIENCE

Recent headlines declare that there is a crisis in higher education in the United States (Hacker and Dreyfus 2011, Taylor 2010). We hear of deep cuts in public funding of higher education, challenges to collective bargaining of faculty, a steep reduction of tenure track faculty among the professoriate, escalating debt for students, poor learning outcomes (Arum and Roksa 2011), and high unemployment for graduates. Only 44.7 percent of students in the graduating class of 2011 held jobs that even needed an undergraduate degree—students with majors such as education, teaching, and engineering fared best in finding jobs that used their skills to continue to work in their area of specialization (Rampell 2011). Similarly, there has been a steady decline in graduate education, and in the numbers of those who go on to join the professoriate. "We are producing too many PhDs," says Mark C. Taylor (2011). The proportion of people with science Ph.D.s who get tenured academic positions in the sciences has declined steadily. The problem is most acute in the life sciences, where only 15 percent find a tenure track position after six years of a Ph.D. in contrast to 55 percent in 1973 (Cyronoski et al. 2011). Private industry does not have enough positions that require a doctorate to absorb the surplus. Some researchers argue that there is an overproduction of Ph.D.s and that it is "scandalous" that politicians and educators continue to speak of a Ph.D. shortage. Yet, as we saw in the previous two chapters, graduate education is geared to a future in academia and academic research was the top career choice for Ph.D. students in 2010; most Ph.D. programs train students specifically for such a future (Cyronoski et al. 2011). Given this overproduction, we need to reassess academe (Taylor 2011).

Predictions of an upcoming boom of jobs in STEM fields have persisted for decades, but as a share of doctoral-level employment opportunities across all sectors, tenure track academic jobs have shrunk dramatically—from 75 percent of the professoriate in the 1970s to an average of 25 percent in recent years (Bousquet 2008). Academic employment opportunities in short-term contract faculty positions with far less job and economic security have increased; women are overrepresented in such positions (Feldman and Turnley 2004). Similarly, in the private sector outsourcing and globalization have created significant wage reductions. Women are concentrated in lower-paying technical jobs rather than in positions as leaders or heads of research labs and departments. It is rather remarkable that despite these shrinking opportunities, there is a continuing effort to recruit women to the ranks of those awarded Ph.D.s. The overproduction of women Ph.D.s in a dwindling job market creates an overqualified but underpaid workforce. In this context, the relentless pursuit of WISE programs is astonishing. There is little discussion or preparation for this reality in WISE-related initiatives. As a result, WISE-related initiatives it would seem are complicit in the creation of a highly skilled but underpaid, underappreciated, underrecognized, and uncertain (largely female) workforce.

THE LIMITED MODELS OF DEFICIT VERSUS DIFFERENCE

Why are women less likely to stay in STEM fields and why are they less likely to be successful in scientific careers (Sonnert and Holton 1996)? Two competing models predominate in analyses of gender differences in educational and career outcomes in STEM disciplines (Sonnert and Holton 1995, Barbercheck 2001). The deficit model assumes that men and women are similar and are motivated by the same goals and aspirations. What accounts for the disparity is that women are disadvantaged by structural obstacles, lower status positions, lower pay, and limited access to resources and reward networks (Sax 2001). Thanks to the long and persistent activism of feminists, many of the formal barriers have been removed, but many subtle informal barriers remain. Those using a deficit model perspective focus on removing the barriers and training women to overcome them. The second model, the difference model, assumes that women and men have inherently different preferences, aspirations, and goals—either through genetics, socialization, or gendered cultural values (Sonnert and Holton 1995). From this perspective, attitudes about science define it as a male domain and, as a result, rigid gender roles discourage girls and women from expressing and developing their interests in science (Valian 1999, Barbercheck 2001). Women are socialized into attitudes and behaviors that deemphasize qualities such as competitiveness, ambition, and aggressiveness, which are central to science today; thus women are disadvantaged because they cannot conform to gender norms as scientists. Those using the difference model argue that rather than training women into a "male" model of science, we need to engage in a reform of science to accommodate a more diverse range of styles, behaviors, and epistemologies (Malcolm 1999, Barbercheck 2001). Using the same arguments, other scholars counter that women's difference explains their continuing marginalization. A well-publicized study by Ceci and Williams (2011) argued that the glass ceiling has been broken, that discrimination no longer existed, and that women "chose" the life in science they wanted (Zakaib 2011). In the difference model, the reason for gender disparities lies squarely in women's differential abilities or aspirations, desires, and goals.

It is unfortunate that for twenty years these two models that take an either/or approach have dominated frameworks for understanding the underrepresentation of women in the sciences. Grounded in equity feminism, the focus of both approaches is on increasing the numbers of "bodies" of women. The deficit model argues for a focus on overcoming structural barriers for girls and women. The actual structure of scientific education and training remains unscathed—rather, the focus has been on getting women to excel *within* that structure. The difference model focuses on developing a more diverse and plural institution of science, but draws on essentialist assumptions about what girls and women are taught, are good at, and enjoy. Again, the main institution of science remains uninterrogated and marked as "scientific," while essentially remaining male. Interventions within the difference model open up new spaces for the feminine and "female-friendly" science, as an accommodation to girls and women, often misunderstood as feminist efforts. Such ideas of difference mark female researchers and femininity as "visibly different" and male researchers and masculine epistemic subjectivity as invisible and unmarked (Harding 1991, Garforth and Kerr 2009). As Helen Longino (1989) eloquently argues, the conflation of feminist and feminine is not useful. Science needs the whole repertoire of human potential, not a pigeonholed vision of two mutually exclusive ways of being—either masculine or feminine.

What is unfortunate in both models is that mainstream science as a gendered institution remains entirely unexamined (Rolin 2004). The very rules of the game in the history of science are grounded in a history of a racialized heteronormative masculinity that precludes the participation of women; feminine science is a contradiction in terms. Furthermore, the seemingly benign vocabulary of "difference" obfuscates deeper claims of essential sex differences. Feminists have pointed out that the relegation of "feminine" and "masculine" traits is neither innocent nor accidental. When marginalized groups are marked as "different," this difference translates into "deficient" in the language of power (Fausto-Sterling 1985, Shiebinger 1989, Hubbard 1990). For example, women may be celebrated as having great "nurturing" talent; unsurprisingly, jobs that require nurturing talent are less prestigious and pay poorly. Furthermore, such "feminine" talents are rarely heralded as important to science and the scientific temperament. Rather than reinscribing traditional gender roles of masculine and feminine, feminism must be a project about troubling gender, about elaborating how gender might be done as well as undone and redone in new and radical ways or even not at all (Morland 2011).

WOMEN IN THE SCIENCES AND THE EMERGENCE OF THE NEOLIBERAL UNIVERSITY SYSTEM

Over the past two decades, scholars have noted deep erosion in support for public higher education. With progressive cuts to public education, many see this as an ideological turn in universities and the emergence of an increasingly neoliberal university (Slaughter and Leslie 2007, Slaughter and Rhoades, 2000, 2004, Aihwa Ong 2006, Wendy Brown 2011). As Aihwa Ong (2006: 140) argues, there is a double movement in U.S. higher education: "a shift from a national to a transnational space for producing knowledgeable subjects, and a shift from a focus on political liberalism and multicultural diversity at home to one on neoliberalism and borderless entrepreneurial subjects abroad." With growing neoliberalism across the globe, public higher education has increasingly put markets as central to social value. Universities have becomes sites for private entrepreneurship, and technoscience is increasingly the central focus of the university. The neoliberal university thus stresses training students close to the technoscience core of knowledge economies, embracing research that creates high-tech products and processes for private capital, and preparing the future workers for a global technoscience economy (Slaughter and Rhoades 2004).

This neoliberal turn has shifted the focus from the mission of higher education for the benefit of all citizens, to a focus on business and privatization of public goods and services.

WISE projects and programs have continued to thrive through the neoliberal turn and are increasingly well integrated into the neoliberal university. They have accomplished this by adopting the language of economic efficiencies, evaluation, and accountability as their universities move toward market/business models of organization and a rhetoric of transparency and public accountability (Deem and Morley 2005, Garforth and Kerr 2009). Programs are tasked with identifying "high-quality" women and people of color who show scholarly distinction and talent, and working to promote them (Iverson 2007). Program annual reports, evaluations, publications, and activities are now available via program websites that celebrate program innovations as opening new horizons for women in science. These are accomplished either through mentoring women to overcome their "deficit" in training or in accommodating women's "difference" in the evaluative logics of the university. In this trend, the language of equity is accompanied by new political and economic language. Programs tout both economistic (arguments about efficiency and the importance of "tapping" into hidden and unused talent) and democratic (liberal sentiments of equal opportunity and inclusion) goals.

Garforth and Kerr (2009) argue that these changes in language reposition discussions about inequalities through a rhetoric of social inclusion that rests on the logic of market economics: how we can produce more women innovators and entrepreneurs and women's startups, innovate in producing female-friendly toys and computer games, and enable women scientists to produce more papers, grants, awards, and patents? Thus left and right, labor and capital, can be brought together in this rhetoric. These two are presumed but not explained. Indeed, scientific practices are recouched in new terms. For example, Louise Morley (1999) argues that the new quality audit produces new forms of macho competitive individualist masculinities with an emphasis on outputs and targets, while women are made responsible for communal caretaking and the domestic elements of academic labor such as teaching and administration. We are thus moving from an old "collegial fraternity" to a new culture of "technocratic patriarchy" (Hearn 2001). To be sure, some women have benefited from these new managerial practices since celebrated "excellent" ones thrive, but these practices also have had profound effects on the ways in which gender inequality has been reproduced, reinforced, and reinvented through gendered employment insecurities. The successful woman scientist is one who exemplifies a new model of efficiency, producing the right number of publications and receiving the requisite number of grants and awards. But this is not true of the vast majority of women. Indeed, women are overrepresented in teaching faculty and teaching colleges. Women and faculty from marginalized groups are over-represented in administrative tasks as they get recruited into various committees in the name of diversity. The rules of the new "audit" of who is productive, who receives "merit" money, and who is released from teaching reinscribe gendered differences that have been sutured into the underlying assumptions and logic of modern organizations. The discourse about discrete barriers or pipeline leaks severely misidentifies the problem. We need to recognize that the very logic of academic careers is gendered (Garforth and Kerr 2009).

Acker (2000), and Meyerson and Kolb (2000) argue that while the rhetoric of equal opportunity, equality, and fairness may continue, any work that is not in an organization's interest will be resisted. Demands of individual perfor-mativity always overshadow economic concerns, so financial implications are always subordinated to gender initiatives. As Garforth and Kerr (2009: 397) summarize:

> The dominant de-individualized economic discourse of women as resource, the reiteration of the linear career path, the dual rhetorics of institutional and policy progress, the re-inscription of domesticity and subjectivity onto visible women, and the proliferation of gender audit and experts that make up the package of "equality works" cloaks male researchers and the masculin-ity of science in its traditional robes of objectivity, neutrality, and normality.

Indeed, if science and academia itself is gendered, if the very calculus and logic is gendered, attempts to include women in this logic without transforming it merely reinscribes a gendered logic. Those men and women who fit into the neoliberal logic indeed do well, rising in the hierarchy and being richly re-warded. The rise of these successful women of "excellence" provides an insidi-ous logic in privatizing the "failure" or "choice" of others who do not or choose not to embody such narrow definitions of "excellence." This fails to challenge the structure of advantage and the gendering of science and the division of la-bor (Wajcman 1991). Recent innovations in women in science initiatives as a result reinscribe gendered ideas that always put the burden on women and the housekeeping of administration, helping the soul of the university and national science initiatives. The deeply embedded rhetoric of nation and nationalism is striking (Garforth and Kerr 2009).

IMPOVERISHED MODEL OF FEMINISM

The idea of feminism in the contemporary United States provokes both back-lash as a radical agenda and a "special interest" political perspective, as well as an idea that has been simultaneously and thoroughly mainstreamed. Equity is a case in point. Polls repeatedly show that the majority of Americans believe in gender equity, and indeed bringing gender equity to the world has been a reason to go to war! Yet equity has been thoroughly critiqued within feminist studies as reinforcing ideas of femininity and feminine difference rather than challenging the gender binary itself (Garforth and Kerr 2009).

The goals of equity, while laudable, have severe limitations. Drawing on the second wave of feminism, equity efforts have attempted to give women equal access to science. As feminists have demonstrated, however, equity efforts have not caught up with personal lives. The whole enterprise of programs for women and minorities is predicated on women and minorities as "always already" victims of oppression (Iverson 2007). Efforts of programs are focused on recruiting and retaining them. Yet these efforts are rarely in tune with the realities of gendered lives outside science. Women continue to be burdened by the "second shift" and with mainstream ideologies of femininity. Indeed, in order for science or mainstream culture to be "de-gendered," we need to concurrently "re-gender" relations between men and women (Lorber 2000). These efforts have been slow. An intransigent scientific culture coupled with a feminism that has reinforced essentialist ideas have been less than ideal in transforming lives for women or science. It should come as no surprise that faculty least likely to use mechanisms such as stopping tenure clocks for pregnancy are women faculty in STEM disciplines. Moving beyond the equity model of overcoming overt discrimination has been exceedingly difficult. Many women scientists have themselves been reluctant to embrace strategies that highlight their gender or identity categories. The exclusive focus on equity has thus been a limiting strategy in addressing the underrepresentation of women in science.

Learning to Count Past Two: Shifting Our Focus from Gender to Science

Alongside concerns about the overproduction of Ph.D.s, there is a crisis of confidence about science and engineering education in the United States. Demographics of student populations show increases in international and "foreign" students (Marklein 2012, Matthews 2010). There are similar increases in working international and "foreign" scientists and engineers (National Science Foundation 2012, Matthews 2010). It is argued that we are failing to train more "native" U.S. scientists and engineers (Bennett 2012). The United States is losing its competitive edge (Malone 2012) to European countries, Japan, and, in more recent years, China and India. Here, some of the literature moves into xenophobic territory, bemoaning the rise in foreign-born scientists and students rather than highlighting the significant disinvestment of public education in the United States and the changing economic contexts that have shifted the flows of labor. Two sets of factors propel these arguments. First, for more than two decades U.S. schoolchildren have consistently performed poorly in science and math tests when compared to students in the rest of the world. This educational deficit is not only in science and mathematics but is systemic and includes reading and writing. As the U.S. secretary of education said, "we're being out-educated" (Kobert 2011: 72). Second, there is much anxiety about the scientific workforce. Fewer U.S. citizens are opting for science and math careers, and this deficit is being filled by a greater reliance on foreign students and foreign-trained workers (Cyranoski et al. 2011, Nelson 2007). What was once dubbed the "brain drain" of a talented pool of foreign scientists and engineers who flocked to the United States from other countries has now been reversed. Increasingly, foreign students with U.S. degrees are leaving the United States after their training and returning to their homelands. In response, Congress and U.S. policy have sought to find ways to keep this talent in the United States. Again, a narrow model of equity in women in science and engineering seems detached from the complex political and demographic practices that are fueling contemporary flows of labor and international capital.

It is in within these larger shifts in national and global contexts that we need to evaluate WISE initiatives. WISE programs have been incredibly important in opening and promoting openings for women in science and engineering. I do not want to understate these efforts or the striking shift in numbers they have engendered. At the same time, the strategies have occasionally elided with larger political and economic shifts that are not always progressive. An exclusive focus on equity has led to a focus on the numbers of women and students of color instead of science. Feminist scholars have spent a great deal of time studying such patterns but there is little engagement with this literature in many WISE-related initiatives. Instead, sex, gender, and race are talked about as stable, ahistorical categories. What if we take the question of variation seriously and attend to our changing understandings of variation, diversity, and difference?

Studies of diversity, rather than questioning racial or other differences, often "mirror" and reinforce them (Baez 2004). For example, as Omi and Winant argue, identity categories like race do not have content by themselves. These categories emerged at particular historical moments and were mobilized toward clear political and economic ends. Race is a "formation" through the complex and shifting sets of political projects that organize human bodies and social structures in the aid of particular agendas (Omi and Winant 1994). The challenge then is to explore how and why our discourses of difference organize the category of difference within particular configurations of power and how they shape individual experiences and attitudes (Baez 2004). Gender and other identity categories are not qualities we "have" or possess but rather what we learn to "do"—it is an active process of construction in relation to historical, political, and economic contexts (Morland 2011).

These are also relational categories, always co-constructed with ideas of race, class, sexuality, and nation. In her classic essay nearly thirty years ago, "How Gender Matters, or, Why It's So Hard for Us to Count Past Two," Evelyn Fox Keller (1992) warned us about the slippage between sex and gender. She recounts that when she told people that she worked on "gender and science," she was usually asked, "So what have you found out about women in the sciences?" She reminded us that our inability to move beyond the binary frames of male/female elides sex and gender where "gender" often reinscribes ideas of femininity and women. And indeed the WISE literature is testament that we have been constantly and consistently caught up in a binary world. Thirty years since the emergence of the early feminist critiques of science, WISE initiatives continue to extol the importance, virtues, and joys of women and femininity while the mainstream culture of science has been left largely uncontested except in making accommodations to marginalized groups. The onus has been on marginalized groups making their case and fighting for inclusion rather than an onus on the scientific enterprise being held responsible for its exclusions.

While many researchers have explored the attrition of women from science, there has been limited work on understanding the *problem* of the underrepresentation of women in sciences (Gonsalves 2011). To redefine the problem of women in science, we need to decenter the category of "woman" and instead shift the center of the problem to that of science. It is science rather than women that needs to be interrogated. Rather than merely accommodating women's different biologies, we need to interrogate the narrow frameworks of "good scientists" and "good science" that continue to be valorized within scientific culture. As the previous two chapters demonstrate, individual scientists are limited by definitive yet narrow notions of what it means to be a good scientist. By excavating the histories that have shaped the narrow vision of scientists and science, we open up the possibilities of new and more inclusive futures. We may yet learn how to count past two.

Feminist efforts for women in the sciences must move beyond just increasing the numbers of women in science. Recent efforts of women in science initiatives that buy into promoting women into narrow definitions of efficiency and productivity elide with eugenic scripts of decreasing rather than increasing variation. Instead, a feminist project must include exploring and transforming scientific culture to include not only different bodies but also different visions and cultures, as well as different epistemologies, methodologies, and methods. The project of bringing more women and people of color into science must occur in concert with examining the knowledge the same academy produces about these groups. I am constantly astonished as my students in the sciences—who have been recruited in an effort to diversify science—tell me about the problematic work on biological differences (sex, race, sexuality, ability, nation) that are presented in their biology classes as "truths." What does it mean to recruit a group into an enterprise that simultaneously teaches them about their own biological inferiority? As parts I and II of this book have argued, scientific knowledge production is deeply implicated in its cultures and practitioners. Feminist transformation of the sciences has not embraced the radical potentials of its own critiques. We need to move beyond the frame of women *in* the sciences to embrace feminist analyses of science in its holistic frames. While women in science initiatives have clothed the emperor in new cloaks of femininity, science, like the emperor, is still quite naked. As the story about the emperor and his clothes go, we don't need to reclothe the emperor in feminine or feminist garb. We need to end monarchies!

New Cartographies of Variation

The Future of Feminist Science Studies

> I wish for all this to be marked on my body when I am
> dead. I believe in such cartography—to be marked by
> nature, not just to label ourselves on a map like the names
> of rich men and women on buildings. We are communal
> histories, communal books. We are not owned or
> monogamous in our taste or experience.
>
> —Michael Ondaatje, *The English Patient*

> If you want to build a ship don't gather people together
> to collect wood and don't assign task and work but rather
> teach them to long for the endless immensity of the sea.
>
> —Antoine de Saint-Exupéry, *The Wisdom of the Sands*

During the summer I was revising this manuscript, I decided to grow morning glories in my garden. An obeisance of sorts to this glorious creature that
had captured my imagination two decades ago and has ever since made me
long for the immensity of its naturecultural possibilities. It has been a glorious journey. Among the bewildering varieties now available, my favorite is the
Japanese morning glory, *Ipomoea nil,* Imperial Star of India! As I ponder the
flowers each morning, the naturecultural genealogies of morning glories I have
recently discovered come tumbling into my mind. I contemplate the genus
with its ancient origins in Pangaea (Mann 2011), migrating with the moving
continents, domesticated, spreading globally through colonial expansion. It is
at once a beloved cultivar and a noxious weed, technologically manipulated,
commercially ebullient, and named with imperial pride. What a story! Morning

glories, through their extraordinary naturecultural genealogies, their geographies mapped to global travels of colonialism and commerce, and, like their thigmotropic tendrils, their biographies woven into so many lives, including mine, embody this book on the question of variation.

When I started exploring the genealogies of the history of invasive species and the question of women in the sciences, similar complex and global naturecultural histories came tumbling out as well. The world, it would seem, and its many naturecultural objects are deeply interconnected and intertwined in their myriad histories. For these objects, and others in the world, no solitary, narrow history will do. And yet our contemporary cartography of knowledge maps this immense interconnected world into an impoverished, narrow, fragmented, hyperspecialized academia, where knowledge has been vivisected into narrow disciplinary and subdisciplinary formations. Interdisciplinarity today is most heralded in the interdisciplinary sciences that remain contained within the natural sciences, firmly ensconced in uncontested epistemologies and ontologies. Naturecultural histories and genealogies suggest that we need new cartographies of study. We need to replace these insular and narrowly focused areas of study with communal histories and communal storytelling. This to me is the profound gift the feminist studies of science has given me—a move from the myopic view of a morning glory field as a controlled experimental object to the expansive vision of a global traveler par excellence, its migrations and evolutions intertwined in a global history. How can I be bound by the myopia of a discipline when feminist studies of science has handed me the world and set me free?

Inspired by my morning glory philosophies, this book explores what is lost when parochial commitments to conventional disciplinary hyperspecialization prevail. I am making an argument for reinventing the ways in which we practice science and feminism, for reimagining nature and culture as naturecultures. Beginning with a central concept in evolutionary biology, variation, I have argued that this biological concept has been deeply intertwined with cultural ideas about diversity and difference since its very inception. Put starkly, evolutionary theories and models of variation owe their formulation to cultural debates around diversity and difference, culminating in their *eugenic scripts* that have haunted us ever since. This has not been a linear, strategic, or overt process. It is not as though scientists developed their theories, and then society enacted them. Rather, science and society have been co-produced and co-constituted through narratives about culture and nature that contain what can and cannot be said. Ignoring these naturecultural connections, I argue, produces ill-informed science and society, as well as impoverished, ill-informed understandings and models of nature and culture. Naturecultural visions show us that individual disciplines are each imbued with cultural norms and histories while being blind to those influences. Instead of these narrow genealogies of variation, I suggest new cartographies of variation, embedded in global circuits of knowledge.

Throughout the book I have used ghosts and the metaphor of haunting to highlight how a disciplinary academy and its knowledge traditions have consigned a brutal history to the invisible world of ghostly hauntings. In my experience, in a traditional academy, scientists learn little of the humanities and humanists learn little of science. As a young evolutionary biologist, I did not formally study the deep eugenic roots of my field and the cost of the horrendous eugenic policies they enabled. As a young feminist, I did not formally study the biological theory that grounds the histories of eugenics and educational policies about women. Biology was always removed from culture, and culture from biology. Women's studies has not been an innocent bystander in this dynamic, as feminist science studies exists only at the margins. From my naturecultural perch, as I pulled down the walls that separated the worlds of natures and cultures, the ghosts emerged, and through their stories I came to understand how thoroughly I had been educated in ignorance.

The metaphor of ghosts comes from Bollywood and Indian movies I grew up with. In this genre, ghosts wish to leave the earthly world and go home to the ghostly world but cannot do so because of injustices they have encountered or issues that remain unresolved. Their ghostly spirits linger in the earthly world, seeking justice and resolution. For me, the hauntings of ghosts represent the silenced eugenic history of ecology and evolutionary biology and our consequent refusal to suitably acknowledge the horrors of eugenics. In the refusal to acknowledge this past, the question of variation and its attendant implications surfaces again and again. These are the ghostly hauntings. Contrary to the obligations and responsibilities of higher education, disciplinary configurations, allegiances, and knowledges promote "agnotology," Robert Proctor's wonderful term that denotes the study of culturally induced ignorance or doubt. Disciplinary formations, I argue, obfuscate the inconvenient, avoid the uncomfortable, and promote ignorance about the profoundly powerful insights of interdisciplinary thinking. In studying the parts, nature and culture as ontologically separate zones, we lose sight of the whole, the naturecultures. Disciplines' narrow formulations of meaningful questions and definitions of objects of study preclude examining connections to other spheres. I have come to these reflections on disciplinary knowledge after years of doing discipline-constrained work and feeling the deep limitations of trying to do interdisciplinary work in a disciplinary academy. I have crafted an interdisciplinary intellectual life, and have worked with and through the institutional structures that disciplines accord in the contemporary university.

This book is modest evidence of the possibilities of taking that particular career path; no doubt it will be too adventurous for many and not adventurous enough for some. At least perhaps it will encourage disciplinary transgressions that are less timid than humble, less conventional than imaginative, and less about teams than about genuine collaboration. We are, I believe, paying too high a price in creativity and innovation for our loyalties and allegiances to our disciplinary groundings.

In particular, I elaborate the concept of naturecultural knowledge, that is, the thoroughly entwined, inseparable synergies of the natural world with humanity. I argue that in a truly collaborative academy that recognizes the connections between natures and cultures, scientists and feminist scholars could acknowledge and accept our culpability in the continuing injustices in the world and then join efforts to disrupt them. Ignoring the historical backdrop of eugenics debates dooms scientists (and I speak here specifically of evolutionary biologists as a particularly influential group) to a future as co-conspirators in the production of inequality. The ghosts cannot be chased off unless they are listened to and heard. In tracing the genealogy of variation, I maintain that the same questions about whether diversity is good or bad have fueled long-enduring debates on nature or nurture. The disciplinary boundaries that dictate professional training mask the continuing salience of these debates. Evolutionary biologists would benefit from realizing that these are not just political questions, they are at the core of biology. Humanists, including those in women's studies, would benefit from realizing that these are not just biological questions, they are at the core of our humanity. Nor are these questions unique to evolutionary theory. The same politics of diversity and difference haunted (and still haunt) the geographies of variation. The same questions about diversity emerged as nation-states built imaginary borders and imaginary communities filled with natives and nationalisms, now sustained in contemporary immigration politics and environmentalisms. Finally, perhaps most sadly, the ghosts of eugenics haunt the very lives and biographies of those who would, or have, become scientists. Until science educators recognize that science developed as a world that *by design* excluded women and people of color and third world scientists, until contemporary science educators grasp and disrupt the historical construction of science as a reification of white male intellectual supremacy, scientific training will reproduce and reward traditional conformist models of science and scientists. By examining the contemporary implications of the nineteenth-century eugenics debates, scientists can begin to confront challenges to cultivating genuine diversity. Understanding the meaning and value of variation, diversity, and difference has been and will continue to be the key challenge of our world unless or until the ghosts of eugenics are exposed, appeased, and put to rest. My argument is that silence about the co-construction of nature and culture, in particular, enables the proliferation of sexism, racism, classism, casteism, homophobia, ableism, and other forms of oppressions. What the ghosts want, in classic Bollywood fashion, is to be listened to, understood, acknowledged, recognized, and resolved. Making visible the connection of science and politics, of science and power, of science and culture, is the beginning of such a project.

The book demonstrates the deep and broad implications of taking a naturecultural perspective. Many of these implications challenge the structures, practices, and processes of creating new knowledge in higher education. A naturecultural vision is at the heart of the feminist studies of science and technology, an interdisciplinary field of study that does not neatly fit into any single disciplinary studies area because it embraces variation and thrives on intellectual adventure. In my case, feminist science studies was still relatively new, so I forged a way to my enduring commitments to science by building bridges to and through women's studies and the feminist critiques of science. While interdisciplinary, I found that women's studies, then and now, leans heavily on the humanities and social sciences for most major works; feminist science and technology studies remain relatively marginal. Within biology, except for the question of women in the sciences, feminist scholarship remains remote and irrelevant. In this book, I have made a plea for the critical importance of bringing these fields into conversation with each other. The stakes are high. The future of the Planet, its ecologies and evolutions, the futures of humanity, its inequalities and injustices, are all just too urgent for us to remain silent in our disciplinary silos. A feminist studies of science and technology—robustly interdisciplinary and supported by the humanities, social sciences, and the natural sciences—offers an antidote to the willful agnotology that disciplinary formations have created. It is this space beyond disciplines, freed from disciplinary shackles of rigor, which I imagine and long for. With such a vision and unbridled imagination, forever cognizant and wise to our sexist, racist, classist, and homophobic past, we can chart new academic practices in making possible new genealogies, geographies, and biographies.

But what is the likelihood of this— how do we pull these interventions out from the margins?

Notes

Preface

Parts of the preface are drawn from Subramaniam (1998).

1. I am deeply indebted to my dear friends in graduate school, especially Rebecca Dunn, Peggy Schultz, Jim Bever, and Mary Malik, for their unflagging support and own forays into the politics of science and feminist and critical race scholarship; without them, my own engagement would have been difficult, if not impossible.

2. I am forever indebted to Jean O'Barr, director of the Women's Studies Program when I was a graduate student, and Mary Wyer, who was managing editor of the journal *Signs* while it was housed at my institution. They were instrumental in steering me through the literature in feminist theory and feminist science studies and worked patiently with me.

Introduction. Interdisciplinary Hauntings: The Ghostly World of Naturecultures

1. I am grateful to my colleague Angela Willey for many hours of discussions on the multiple genealogies of feminist science and technology studies. Her astute observations have considerably shaped my historical understanding of the emergence and development of the field.

Chapter One. Thigmatropic Tales: On the Politics and Social Lives of Morning Glories

1. I do realize that the term *eugenics* has historically specific origins. I'm using the term loosely here to indicate how this history translates into modern-day sensibilities.

2. Morning glory flower color variation is well studied. See, for example, Baucom et al. 2011, Barbara A. Brown and Clegg 1984, Epperson 1986, Epperson and Clegg 1983, Fine-

blum and Rausher 2002, Glover et al. 1996, Zufall and Rausher 2003, Shu-Mei Chang and Rausher 1999.

3. In 1911, Wilhelm Johannsen introduced the terms *genotype* and *phenotype* to refute the "transmission view" of genetics, where it was believed that personal qualities could influence the passing down of traits. He introduced these terms as way to signal the shift from an old science of "heredity" to a new science of "genetics." See Gudding (1996).

4. From a feminist standpoint, the language of genetics is fascinating. Alleles are said to be "dominant" and "recessive," even though these are not binary choices since alleles can be "co-dominant," as in the case of the w locus. While individuals may carry a particular allele, whether this is reflected in the emerging phenotype is not always self-evident—the proportion of individuals that express the phenotype is reflected in the "penetrance" of the allele or mutation. The field is rife with gendered metaphors and imagery.

Chapter Two. A Genealogy of Variation: The Enduring Debate on Human Differences

1. I entered the world of the history of eugenics expecting to find a clear link to the work I was doing. There was no singular history of "variation" that traced the idea through the history of eugenics, however. I also expected a well-researched field but was ill prepared for the sheer volume of history and historiography on eugenics! Eugenics, it would appear, is a discipline of its own. What I imagined would be a footnote grew into chapters 2 and 3 of this book along with a healthy respect for the painstaking work of historians!

2. My readings of history surprised me. What I found were distinct genres—internalist histories, critical histories, controversies in biology, biographies, period histories, and so forth. Each of these is a distinct genre, each rich and vibrant. But the internalist (told from within science) versus externalist (told from outside) are two broad frames that shape the history of science. See Allen (1986) for an internalist/externalist history of eugenics. See Ruse (2005), Grene (1983), Bashford and Levine (2010), and P. Levine and Bashford (2010) for their analyses on shifts in the fields of the history and philosophy of science from internalist to externalist explanations, and a shift from analyses of "pure" science to science as a set of practices located and shaped by particular histories, cultures, and politics.

3. For more on the debates on essentialism and evolutionary biology, see Winsor (2003) and Walsh (2006).

4. Several tenets of the modern synthesis led to debates once the molecular era began (Depew and Weber 1995). For example, questions about the units of selection and whether selective process can act at any and every level of scale continue to generate considerable debate (Depew and Weber 1995, Gould 2002).

5. The "new genetics" usually refers to the growth of genetic research and technology arising out of recombinant DNA in the 1970s.

Chapter Three. Singing the Morning Glory Blues:
A Fictional Science

1. This character is inspired by Prashanta Mahalonobis's description of the philosophy of the Indian mathematician Srinivasa Ramanujam and Ashis Nandy's (1995) own description and analysis. I have, of course, embellished and taken many liberties with their descriptions.

Chapter Four. Alien Nation: A Recent *Biography*

Some parts of Act I were previously published in Subramaniam (2001).

1. Connections between the body as fortress and the nation as fortress in late capitalism can be seen in Martin (1997). For thinking about the persistence of national states in an age of globalization, see Comaroff and Comaroff (2000, 2002).

2. For provocative thoughts about the production of ethnicity and civil society/nation, botanical taxonomies, and immigration policies in the United States, see Moallem and Boal (1999). For some recent work on ethnocentrism and nationalism produced as a certain politics, see Paola Bachetta's article on xenophobia and the Hindu right in India (1999). For national cultures, cultures, and questions of cultural nationalism, see Sahlins (2000). For cultural nationalism and new modes of citizenship, see Aihwa Ong (1999).

3. While the distribution of political and religious beliefs in science may not parallel mainstream culture, multiple and diverse political and ideological beliefs are indeed represented.

4. The campaigns are, of course, not by the same groups. Many ecologists have expressed reservations about genetically modified food. My point is about the rhetoric that circulates in the mainstream United States.

5. Anna Tsing (1994) makes a similar point in her analysis of native and exotic bees.

Chapter Five. My Experiments with Truth:
Studying the Biology of Invasions

I offer this chapter title with due apologies to M. K. Gandhi's *The Story of My Experiments with Truth*. As an experimental biologist, I could not resist it!

This work was supported by an NSF grant to the author for "Impact of Soil Communities on Invasive Plant Species in Southern California" (NSF 0075072)

1. Jim Bever and Peggy Schultz were at the University of California, Irvine, when this project began and subsequently moved to Indiana University.

2. The native species were *Lotus purshianus, Salvia apiana, Nassella pulchra, Achillea millefolium, Galium angustifolium,* and *Isocoma menziesii,* and the naturalized species were *Rumex crispus, Bromus diandrus, Lactuca serriola, Medicago polymorpha, Lolium multiflorum, Marrubium vulgare, Salsola tragus,* and *Sonchus asper.*

3. A survey of how biologists categorize various species would be fascinating.

Chapter Six. Aliens of the World Unite!
A Meditation on Belonging in a Multispecies World

1. This pattern is not universal and there are many exceptions. Not all invasives are named with their geographical origins. Why and how this happens is a fascinating future project.

2. On this issue, see also Pollan (1994) and Olwig (2003).

3. For a discussion of Denmark, see Olwig (2003), and of Britain, see Smout (2003).

4. Some do contest such a reading of Jensen; see Grese (2011).

5. The cookbook they have produced is Wong and Leroux (2012).

Chapter Seven. Through the Prism of Objectivity:
Dispersions of Identity, Culture, Science

1. I was deeply influenced by the autobiographical narratives of E. Keller (1977), Weinstein (1977), and Hammonds (1993).

2. I do recognize that the experience of cultural isolation is field dependent. Fields such as engineering or business have a large number of South Asian students who came from elite and prestigious universities in South Asia, and their experiences may be very different.

3. This also pervades the workforce in this country with respect to salary. It is a common joke among Indians here that you can ask another Indian how much she or he earns but not about his or her sex life. Americans will tell you all about their sex life but salaries are too personal!

Chapter Eight. Resistance Is Futile! You Will Be Assimilated:
Gender and the Making of Scientists

This work was supported by an NSF grant to the author for "Breaking the Silences: A Faculty-Student Action Project for Graduate Women in Science and Mathematics." (HRD9553439)

1. Perhaps one of the most interesting species on *Star Trek: The Next Generation*, the Borg have captured the imagination of many "trekkies." A number of websites have proliferated, the most interesting of which is the Borg Institute of Technology (BIT) with the maxim "Graduation is futile!" (previously http://grove.ufl.edu/~locutus/Bit/bit .html, now http://www.skippypodar.net/Bit/.

2. I am deeply indebted to Ann Gerber and Laura Briggs, who were co-facilitators with me during these sessions. Their observations and rich insights have profoundly shaped the interpretation and analysis of this work.

3. It is also important to note that there are strict hierarchies within faculty, especially within the same department, and faculty themselves are sometimes disempowered by the structures of academe, factors often not always visible to graduate students.

Chapter Nine. The Emperor's New Clothes:
Revisiting the Question of Women *in* the Sciences

1. In connecting scientific knowledge to structures of power, the feminist literature suggests systematic discrimination and exclusion of the "nonnormative"—women scientists, scientists of color, working-class scientists, third world scientists, queer scientists, non-able-bodied scientists. "Women," after all, are not a universal group, and there are considerable differences among women within and between countries. These differences, however, are not that well studied, and what remains is largely a history of elite white women's efforts to gain entry into the hallways of science.

2. While white women have been the largest beneficiaries of affirmative action in the United States, other underrepresented groups have also pushed for equity measures in STEM disciplines, including groups promoting "minority," queer, disabled, and third world communities.

3. For a discussion of K–12 innovations, see Brotman and Moore (2008), Buck et al. (2012), Hill et al. (2010), Nosek et al. (2009), Scher and O'Reilly (2009), and Valla and Williams (2012). For innovations at the undergraduate level, see M. J. Chang, Cerna, et al. (2008), M. J. Chang, Eagan, et al. (2011), Cheryan and Plaut (2010), Else-Quest et al. (2010), Fox et al. (2011), Sonnert and Fox (2012), Hanson (2004), Hill et al. (2010), Nassar-McMillan et al. (2011), and Sax (2001). For research and programs at the graduate, postgraduate, faculty, and institutional levels, see Jill Adams (2008), Bhatacharjee (2004), Bilimoria and Liang (2011), Brickhouse et al. (2006), Bystydzienski and Bird (2006), Cech and Blair-Loy (2010), Hill et al. (2010), Gonsalves (2011), Intemann (2009), Leonard (2003), Maria Ong et al. (2011), Robinson (2011), Rosser (1990, 1995, 1997, 2004b), Scantlebury (2010), Schibeinger (1999, 2003), Stewart et al. (2007), and Wylie (2011).

References

Abir-Am, Pnina, and Dorinda Outram, eds. 1987. *Uneasy Careers and Intimate Lives: Women in Science, 1789–1979.* New Brunswick, NJ: Rutgers University Press.

Acker, Joan. 2000. "Gendered Contradictions in Organizational Equity Projects." *Organization* 7 (4): 625–32.

Action News 6, Philadelphia. 2002. "Killer Chinese Fish Surfaces in Maryland," July 10.

Adams, Jill. 2008, February 8. "Nurturing Women Scientists." *Science* 319: 831–36.

Adams, Mark. 2000. "Last Judgment: The Visionary Biology of J. B. S. Haldane." *Journal of the History of Biology* 33: 457–91.

Adams, Vincanne, and Stacy Leigh Pigg. 2005. *Sex in Development: Science, Sexuality and Morality in Global Perspective.* Durham, NC: Duke University Press.

Ahmed, S. 2008. "Open Forum Imaginary Prohibitions: Some Preliminary Remarks on the Founding Gestures of the 'New Materialism.'" *European Journal of Women's Studies* 15 (1): 23–39.

———. 2012. *On Being Included: Racism and Diversity in Institutional Life.* Durham, NC: Duke University Press.

Aisenberg, Nadya, and Mona Harrington. 1988. *Women of Academe: Outsiders in the Sacred Grove.* Amherst: University of Massachusetts Press.

Alaimo, S., and S. Hekman, eds. 2008. *Material Feminisms.* Bloomington: Indiana University Press.

Allen, Garland. 1986. "The Eugenics Record Office at Cold Spring Harbor, 1910–1940: An Essay in Institutional History." *Osiris* (Second Series) 2: 225–64.

———. 1996. "Science Misapplied: The Eugenics Age Revisited." *Technology Review* 99 (6): 22–31.

———. 2001. "Is a New Eugenics Afoot?" *Science* 294 (5540): 59–61. DOI: 10.1126/science.1066325.

Alonso, W. 1995. "Citizenship, Nationality and Other Identities." *Journal of International Affairs* 48: 585–99.

Alsop, Peter. 2009, November. "Invasion of the Longhorn Beetles." *Smithsonian Magazine,* http://www.smithsonianmag.com/science-nature/Invasion-of-the-Longhorns.html.

Alyokin, Andrei. 2011. "Non-Natives: Put Biodiversity at Risk." *Nature* 475: 36.

Anandhi, S. 1998. "Reproductive Bodies and Regulated Sexuality: Birth Control Debates in Early Twentieth-Century Tamilnadu." In *A Question of Silence: Sexuality Economics in Modern India,* edited by Mary E. John and Janaki Nair. Delhi: Kali for Women.

Anderson, Warwick. 2006. *The Cultivation of Whiteness: Science, Health, and Racial Destiny in Australia.* Durham, NC: Duke University Press.

———. 2008. *Colonial Pathologies: American Tropical Medicine, Race, and Hygiene in the Philippines.* Durham, NC: Duke University Press, 2006.

Arum, Richard, and Josipa Roksa. 2011. "Your So-Called Education," *The New York Times,* May 14.

Auld, B. A., and R. W. Medd. 1987. *Weeds: An Illustrated Botanical Guide to the Weeds of Australia.* Melbourne: Inkata.

Babbitt, Bruce. 1998. "Statement by Secretary of the Interior Bruce Babbitt on Invasive Alien Species." Science in Wildland Weed Management Symposium, Denver, CO, April 8, http://www.nps.gov/plants/alien/pubs/bbstat.htm.

Bachetta, Paola. 1999. "When the (Hindu) National Exiles Its Queers." *Social Text* 17 (4): 141–66.

Baez, Benjamin. 2004. "The Study of Diversity: The 'Knowing of Difference' and the Limits of Science." *Journal of Higher Education* 75 (3): 285–306.

Ball, George. 2006. "Border War," *The New York Times,* March 19.

Barad, Karen. 2007. *Meeting the Universe Halfway: Quantum Physics and the Entanglement of Matter and Meaning.* Durham, NC: Duke University Press.

Barbercheck, Mary. 2001. "Science, Sex, and Stereotypical Images in Scientific Advertising." In *Women, Science, and Technology: A Reader in Feminist Science Studies,* edited by Mary Wyer, Mary Barbercheck, Donna Giesman, Hatice Orun Ozturk, and Marta Wayne. New York: Routledge.

Barkley, T. M., ed. 1986. *Flora of the Great Plains.* Lawrence: University Press of Kansas.

Barlow, Connie, ed. 1995. *Evolution Extended: Biological Debates on the Meaning of Life.* Cambridge, MA: MIT Press.

Barnes, Jeff. 2002. "Juvenile Frankenfish Raise the Odds of Alien Invasion," *Washington Times,* July 13.

Baron-Cohen, Simon. 2003. *The Essential Difference: Men, Women, and the Extreme Male Brain.* London: Allen Lane.

Barres, Ben. A. 2006, July 13. "Does Gender Matter?" *Nature* 442: 133–36.

Barringer, Felicity. 2012. "New Rules Seek to Prevent Invasive Stowaways," *The New York Times,* April 7.

Barton, Angela Calabrese. 2001. "Capitalism, Critical Pedagogy and Urban Science Education: An Interview with Peter McLaren." *Journal of Research in Science Teaching* 38 (8): 847–59.

Bashford, Alison. 2010. "Internationalism, Cosmopolitanism, and Eugenics." In *The Oxford Handbook of the History of Eugenics,* edited by Alison Bashford and Philippa Levine, 154–72. Oxford: Oxford University Press.

Bashford, Alison, and Philippa Levine, eds. 2010. *The Oxford Handbook of the History of Eugenics.* Oxford: Oxford University Press.

Baskin, Y. 2002. *A Plague of Rats and Rubbervines: The Growing Threat of Species Invasions.* Washington, DC: Island.

Batten, Katherine M., Kate M. Scow, and Erin Espeland. 2008. "Soil Microbial Community Associated with an Invasive Grass Differentially Impacts Native Plant Performance." *Microbial Ecology* 55: 220–28.

Baty, Phil. 2010. "What's Sex Got to Do with It?" *Times Higher Education,* September 30.

Baucom, R. S., S-M. Chang, J. M. Kniskern, M. D. Rausher, and J. R. Stinchcombe. 2011. "Morning Glory as a Powerful Model in Ecological Genomics: Tracing Adaptation through Both Natural and Artificial Selection." *Heredity* 107: 377–85.

Bean, A. R. 2007. "A New System for Determining Which Plant Species are Indigenous in Australia." *Australian Systematic Botany* 20: 1–43.

Beatty, John. 1987. "Weighing the Risks: Stalemate in the Classical/Balance Controversy." *Journal of the History of Biology* 20 (3): 289–319.

Begley, Sharon. 2006. "He, Once a She, Offers Own View on Science Spat," *Wall Street Journal,* July 13, B1.

Benjamin, Rich. 2012. "The Gated Community Mentality," *New York Times,* March 29, 2012.

Bennett, William J. 2012. "U.S. Lag in Science, Math a Disaster in the Making," February 9, http://www.cnn.com/2012/02/09/opinion/bennett-stem-education.

Beoku-Betts, Josephine. 2004. "African Women Pursuing Graduate Studies in the Sciences: Racism, Gender Bias, and Third World Marginality." *NWSA Journal* 16 (1): 116–35.

Bever, James D. 2003. "Soil Community Dynamics and the Coexistence of Competitors: Conceptual Frameworks and Empirical Tests." *New Phytologist* 157: 465–73.

Bhattacharjee, Y. 2004. "Family Matters: Stopping Tenure Clock May Not Be Enough." *Science* 306: 2031–33.

Bilimoria, Diana, and Xiangfen Liang. 2011. *Gender Equity in Science and Engineering: Advancing Change in Higher Education.* London: Taylor & Francis.

Birke, Linda. 1999. *Feminism and the Biological Body.* Edinburgh: Edinburgh University Press.

Blair, Robert B. 2008. "Creating a Homogenous Avifauna." In *Urban Ecology: An International Perspective on the Interaction between Humans and Nature,* edited by John Marzluff, 405–24. New York: Springer.

Bleier, Ruth. 1984. *Science and Gender: A Critique of Biology and Its Theories on Women.* New York: Pergamon.

Bliss, Catherine. 2012. *Race Decoded: The Genomic Fight for Social Justice.* Stanford, CA: Stanford University Press.

Bombardieri, Marcella. 2005. "Harvard Women's Group Rips Summers," *Boston Globe,* January 19, http://www.boston.com/news/education/higher/articles/2005/01/19/harvard_womens_group_rips_summers/?page=full.

Bousquet, Marc. 2008. *How the University Works: Higher Education and the Low Wage Nation.* New York: NYU Press.

Brabazon, Tara. 1998, May. "What's the Story Morning Glory? Perth Glory and the Imagining of Englishness." *Sporting Traditions* 14 (2): 53–66.

Bradshaw, G. A., and M. Bekoff. 2001. "Ecology and Social Responsibility: The Re-embodiment of Science." *Trends in Ecology and Evolution* 16 (8): 460–65.

Brandon, Robert. N. 1990. *Adaptation and Environment.* Princeton, NJ: Princeton University Press.

Brickhouse, Nancy W., Margaret A. Eisenhart, and Karen Tonso. 2006. "Forum Identity Politics in Science and Science Education." *Cultural Studies of Science Education* 1: 309–24.

Briggs, Laura. 2002. *Reproducing Empire: Race, Sex, Science, and U.S. Imperialism in Puerto Rico.* Berkeley: University of California Press.

Bright, Christopher. 1998, November/December. "Alien Threat." *World Watch* 11 (6): 8.

———. 1999. "Invasive Species: Pathogens of Globalization." *Foreign Policy* 116 (51): 14.

Brinckman, Jonathan. 2001. "Creepy Strangler Climbs Oregon's Least-Wanted List," *Oregonian*, February 28, 1A.

Brockway, Lucile. 2002. *Science and Colonial Expansion: The Role of the British Royal Botanic Gardens.* New Haven, CT: Yale University Press.

Brotman, Jennie S., and Felicia M. Moore. 2008. "Girls and Science: A Review of Four Themes in the Science Education Literature." *Journal of Research in Science Teaching* 45 (9): 971–1002.

Brown, Barbara A., and Michael T. Clegg. 1984, July. "Influence of Flower Color Polymorphism on Genetic Transmission in a Natural Population of the Common Morning Glory, Ipomoea purpurea." *Evolution* 38 (4): 796–803.

Brown, J. H., and D. F. Sax. 2004. "An Essay on Some Topics Concerning Invasive Species." *Australian Ecology* 29: 530–36.

———. 2005. "Biological Invasions and Scientific Objectivity: Reply to Cassey et al." *Australian Ecology* 30: 481–83.

Brown, Wendy. 2011, Summer. "Neoliberalized Knowledge." *History of the Present* 1 (1): 113–29.

Brundrett, M. C. 2002. "Coevolution of Roots and Mycorrhizas of Land Plants." *New Phytologist* 154: 275–304.

Buchanan, Gale A., and Earl R. Burns. 1971, September. "Weed Competition in Cotton. I. Sicklepod and Tall Morningglory." *Weed Science* 19 (5): 576–79.

Buck, Gayle A., Nicole M. Beeman-Cadwallader, and Amy E. Trauth-Nare. 2012. "Keeping the Girls Visible in K–12 Science Education Reform Efforts: A Feminist Case Study on Problem Based Learning." *Journal of Women and Minorities in Science* 18 (2): 153–78.

Bug, Amy. 2003. "Has Feminism Changed Physics?" *Signs: Journal of women in culture and society* 28 (3): 881–900.

Burdick, Alan. 2005, May. "The Truth About Invasive Species: How to Stop Worrying and Learn to Love Ecological Intruders." *Discover Magazine* 26 (5).

Burelli, Joan. 2010, July. "Foreign Science and Engineering Students in the United States." InfoBrief, NSF 10–324, Arlington, VA.

Burian, R. M. 1983. "Adaptation." In *Dimensions of Darwinism*, edited by M. Grene, 286–314. Cambridge: Cambridge University Press.

Butler, Judith. 2006. *Gender Trouble: Feminism and the Subversion of Identity*. New York: Routledge.

Bystydzienski, Jill, and Sharon R. Bird, eds. 2006. *Removing Barriers: Women in Academic Science, Technology, Engineering, and Mathematics*. Bloomington: Indiana University Press.

Calautti, R. I., and J. J. MacIsaac. 2004. "A Natural Terminology to Define 'Invasive' Species." *Diversity and Distributions* 10: 135–41.

Campbell, Nancy. 2009. "Reconstructing Science and Technology Studies: Views from Feminist Standpoint Theory." *Frontiers* 30 (1): 1–29.

Canel, Annie, Ruth Oldenziel, and Karin Zachmann. 2000. *Crossing Boundaries, Building Bridges: Comparing the History of Women Engineers, 1870s–1990s*. Amsterdam: Harwood Academic.

Caplan, Arthur. L., Glenn McGee, and David Magnus. 1999. "What Is Immoral about Eugenics?" *British Medical Journal* 319 (7220): 1284.

Cardozo, Karen, and Banu Subramaniam. 2013, February. "Assembling Asian/American Naturecultures: Orientalism and Invited Invasion." *Journal of Asian American Studies* 16 (1).

Carey, Toni Vogel. 1998. "The Invisible Hand of Natural Selection and Vice Versa." *Biology and Philosophy* 13: 427–42.

Carthey, Alexandra J. R., and Peter B. Banks. 2012, February. "When Does an Alien Become a Native Species? A Vulnerable Native Mammal Recognizes and Responds to Its Long-Term Alien Predator." *PLoS One* 7 (2): 1–4.

CBS Evening News. 2002a. "Freak Fish Found in Two More States," August 3.

———. 2002b. "Wanted Dead: Voracious Walking Fish," July 3, www.cbsnews.com/stories/2002/07/03/eveningnews/main514182.shtmil.

Cech, Erin A., and Mary Blair-Loy. 2010. "Perceiving Glass Ceilings? Meritocratic versus Structural Explanations of Gender Inequality among Women in Science and Technology." *Social Problems* 57 (3): 371–97.

Ceci, Stephen J., and Wendy M. Williams. 2011, February 7. "Understanding Current Causes of Women's Underrepresentation in Science." *Proceedings of the National Academy of Sciences*. DOI 10.1073/pnas.1014871108.

Chaney, Lindsay, and Regina S. Baucom. 2012. "The Evolutionary Potential of Baker's Weediness Traits in the Common Morning Glory, *Ipomoea purpurea* (Convolvulaceae)." *American Journal of Botany* 99 (9): 1524–30.

Chang, Gordon. 2001. *Morning Glory, Evening Shadow: Yamato Ichihashi and His Internment Writings*. Stanford, CA: Stanford University Press.

Chang, Kenneth. 2012. "Bias Persists for Women of Science, a Study Finds." *The New York Times*, September 24.

Chang, M. J., O. Cerna, J. Han, and V. Saenz. 2008. "The Contradictory Roles of Institutional Status in Retaining Underrepresented Minorities in Biomedical and Behavioral Science Majors." *Review of Higher Education* 31: 433–64.

Chang, M. J., M. K. Eagan, M. H. Lin, and S. Hurtado. 2011. "Considering the Impact of Racial Stigmas and Science Identity: Persistence among Biomedical and Behavioral Science Aspirants." *Journal of Higher Education* 82: 564–96.

Chang, Shu-Mei, and Mark D. Rausher. 1999, October. "The Role of Inbreeding Depression in Maintaining the Mixed Mating System of the Common Morning Glory, *Ipomoea purpurea.*" *Evolution* 53 (5): 1366–76.

Chaudhary, Bala, and Margot Griswold. 2001, Spring. *Ecesis: Newsletter of the Society for Ecological Restoration* (SERCAL), http://www.newfieldsrestoration.com/PDFs/Mycorrhizal_Fungi_Ecesis.pdf.

Cheater, Mark. 1992, September/October. "Alien Invasion: They're Green, They're Mean, and They May Be Taking Over a Park or Preserve Near You." *Nature Conservancy* 42 (5): 24–29.

Cheryan, Sapna, and Victoria C. Plaut. 2010. "Explaining Underrepresentation: A Theory of Precluded Interest." *Sex Roles* 63: 475–88.

Chew, Matthew, and Andrew L. Hamilton. 2011. "The Rise and Fall of Biotic Nativeness: A Historical Perspective." In *Fifty Years of Invasion Ecology: The Legacy of Charles Elton*, edited by David M. Richardson. Oxford: Blackwell.

Chew, Matthew, and M. D. Laubichler. 2003. "Natural Enemies—Metaphor or Misconception?" *Science* 301: 52–53.

"Chinese Snakehead Not Maryland's Only Foreign Problem." 2002. *Maryland Daily Record.* July 19, http://thedailyrecord.com/2002/07/19/chinese-snakehead-not-maryland8217s-only-foreign-problem/.

Clegg, Michael T., and Mary L. Durbin. 2000. "Flower Color Variation: A Model for the Experimental Study of Evolution." *Proceedings of the National Academy of Sciences USA* 97 (13): 7016–23.

CNN News Online. 2002. "Wanted: Snakehead," July 12, http://fyi.cnn.com/2002/fyi/news/07/12/news.for.you.

Coase, Ronald. 1998. "The New Institutional Economics." *American Economic Review* 88 (2).

Coates, Peter. 2003. "Editorial Postscript: Naming Strangers in the Landscape." *Landscape Research* 28: 131–37.

———. 2006. *American Perceptions of Immigrant and Invasive Species: Strangers on the Land.* Berkeley: University of California Press.

Cole, J. R. 1979. *Fair Science: Women in the Scientific Community.* New York: Collier Macmillan.

Collins, Patricia Hill. 1999. "Moving beyond Gender: Intersectionality and Scientific Knowledge." In *Revisioning Gender*, edited by Myra M. Ferree, Judith Lorber, and Beth B. Hess, 261–84. Thousand Oaks, CA: Sage.

Comaroff, Jean, and John Comaroff. 2000. "Millennial Capitalism: First Thoughts on Second Coming." *Public Culture* 12 (2): 291–343.

———. 2002. "Alien-Nation: Zombies, Immigrants, and Millennial Capitalism." *South Atlantic Quarterly* 104 (4): 779–805.

Conlon, Michael. 2010. "Flowering Cherry Trees—A Gift from Japan." USDA Foreign Agricultural Service Global Agricultural Information Network, GAIN Report Number: JA0507, http://www.usdajapan.org/en/reports/20100330_Japanese%20Flowering%20Cherry%20Trees.pdf.

Cooke, Kathy. J. 1997. "From Science to Practice, or Practice to Science? Chickens and Eggs in Raymond Pearl's Agricultural Breeding Research, 1907–1916." *Isis* 88: 62–86.

Coronado, Ty. 2006. "LA Immigrant Rights March," *ZNet*, March 31, http://www.zmag.org/content/showarticle.cfm?SectionID=72&ItemID=10022.

Cowan, Ruth Schwartz. 1997. *A Social History of American Technology.* New York: Oxford University Press.

———. 2008. *Heredity and Hope: The Case for Genetic Screening.* Cambridge, MA: Harvard University Press.

Cowherd, Kevin. 2002. "Soon Ghastly Fish Will Walk Off into the Sunset," *Baltimore Sun*, August 8.

CQ Researcher. 2000. "Bio-invasions Spark Concerns." 9 (37): 856.

Cravens, Hamilton. 1978. *The Triumph of Evolution: American Scientists and the Heredity-Environment Controversy, 1900–1940.* Philadelphia: University of Pennsylvania Press.

Creager, Angela, Elizabeth Lunbeck, and Londa Schiebinger, eds. 2001. *Feminism in Twentieth-Century Science, Technology, and Medicine.* Chicago: University of Chicago Press.

Crenshaw, Kimberle. 1991. "Mapping the Margins: Intersectionality, Identity Politics, and Violence against Women of Color." *Stanford Law Review* 43 (6): 1241–99.

Cronon, William. 1996a. "The Trouble with Wilderness; or, Getting Back to the Wrong Nature." In *Uncommon Ground: Rethinking the Human Place in Nature*, edited by William Cronon, 69–90. New York: Norton.

———. 1996b. "Introduction: In Search of Nature." In *Uncommon Ground: Rethinking the Human Place in Nature*, edited by William Cronon, 23–56. New York: Norton.

Crosby, Alfred W. 1986. *Ecological Imperialism: The Biological Expansion of Europe, 900–1900.* New York: Cambridge University Press.

Crow, James. F. 1987. "Dobzhansky and Overdominance." *Journal of the History of Biology* 20 (3): 351–80.

Crowley, R. H., and G. A. Buchanan. 1982. "Variations in Seed Production and Response to Pests of Morningglory (*Ipomoea*) Species and Smallflower Morningglory (*Jacquemontia tamnifolia*)." *Weed Science* 30: 187–90.

Crutzen, P. J., and E. F. Stoerme. 2000. "The Anthropocene." *Global Change Newsletter* 41: 17–18.

Cyranoski, David, Natasha Gilbert, Heidi Ledford, Anjali Nayar, and Mohammed Yahla. 2011. "Education: The PhD Factory." *Nature* 472: 276–79.

Dahl, Linda. 2001. *Morning Glory: A Biography of Mary Lou Williams.* Berkeley: University of California Press.

Darity, William, Jr. 1995. Introduction. In *Economics and Discrimination,* edited by William Darity Jr. Brookfield, VT: Edward Elgar.

Dart, Bob. 2002, July 24. "Invasive 'Walking' Fish Found Across U.S. South." *National Geographic.*

Darwin, Charles. 1922. *Descent of Man and Selection in Relation to Sex.* New York: D. Appleton and Co.

———. 1959. *The Origin of Species by Means of Natural Selection or The Preservation of Favoured Races in the Struggle for Life.* Middlesex: Penguin.

Daston, Lorraine, and Katherine Park. 2001. *Wonders and the Order of Nature, 1150–1750.* Zone.

Davis, Cinda-Sue. 1996. *The Equity Equation: Fostering the Advancement of Women in the Sciences, Mathematics, and Engineering.* San Francisco: Jossey-Bass.

Davis, Kathy. 2008. "Intersectionality as Buzzword: A Sociology of Science Perspective on What Makes a Feminist Theory Successful." *Feminist Theory* 9 (1): 67–85.

Davis, M. A. 2009. *Invasion Biology.* Oxford: Oxford University Press.

Davis, M. A., and K. Thompson. 2000. "Eight Ways to Be a Colonizer; Two Ways to Be an Invader." *Bulletin of the Ecological Society of America* 81: 226–30.

Davis, M. A., et al. 2011a, June. "Don't Judge Species on Their Origins." *Nature* 474: 153–54.

———. 2011b. "Researching Invasive Species 50 years after Elton: A Cautionary Tale." In *Fifty Years of Invasion Ecology: The Legacy of Charles Elton,* edited by David M. Richardson, 269–74. Oxford: Blackwell.

Davis, Noela. 2009. "New Materialism and Feminism's Anti-Biologism: A Response to Sara Ahmed." *European Journal of Women's Studies* 16 (1): 67–80.

Dawkins, Richard. 1986. *The Blind Watchmaker: Why the Evidence of Evolution Reveals a Universe Without Design.* New York: Norton.

De Laey, J. J., S. Ryckaert, and A. Leys. 1985. "The 'Morning Glory' Syndrome." *Ophthalmic Genetics* 5 (1/2): 117–24.

De Laplante, K. 2004. "Toward a More Expansive Conception of Ecological Science." *Biology and Philosophy* 19: 263–81.

Deem, Rosemary, and Louise Morley. 2005. "Negotiating Equity in Higher Education Institutions." Paper presented at the Economic and Social Research Council (ESRC) Seminar on Equalities in Education, University of Cardiff, http://dera.ioe.ac.uk/5886/1/rd10_05.pdf.

Defelice, Michael. 2001. "Tall Morningglory, *Ipomoea purpurea* (L.) Roth—Flower or Foe?" *Weed Technology* 15: 601–6.

Demeritt, D. 1998. "Science, Social Constructivism and Nature." In *Remaking Reality: Nature at the Millennium,* edited by N. Castree and B. Braun, 173–93. London: Routledge.

Department of Natural Resources. n.d. http://dnr.maryland.gov/fisheries/bass/docs/snakehead_contest_20110420.pdf.

Depew, David J. 2010. "Darwinian Controversies: An Historiographical Recounting." *Science and Education* 19: 323–66.

Depew, David, and Bruce Weber. 1995. *Darwinism Evolving—Systems Dynamics and the Genealogy of Natural Selection.* Cambridge, MA: MIT Press.

Derr, P. G., and E. M. McNamara. 2003. "Have You Seen This Fish?" In *Case Studies in Environmental Ethics.* Lanham, MD: Rowman & Littlefield.

Des Jardins, Julie. 2010. *The Madame Curie Complex: The Hidden History of Women in Science.* New York: Feminist Press.

Desrosières, Alain. 1998. *The Politics of Large Numbers: A History of Statistical Reasoning.* Cambridge, MA: Harvard University Press.

Devine, Robert. 1999. *Alien Invasion: America's Battle with Non-native Animals and Species.* Washington, DC: National Geographic Society.

Dietrich, Michael R. 1994. "The Origins of the Neutral Theory of Molecular Evolution." *Journal of the History of Biology* 25: 21–59.

———. 2006. "From Mendel to Molecules: A Brief History of Evolutionary Genetics." In *Evolutionary Genetics: Concepts and Case Studies,* edited by Charles W. Fox and Jason B. Wolf, 3–13. New York: Oxford University Press.

Dietz, T., and P. C. Stern. 1998. "Science, Values, and Biodiversity." *BioScience* 48: 441–44.

Dikötter, Frank. 1998. "Race Culture: Recent Perspectives on the History of Eugenics." *American Historical Review* 103 (2): 467–78.

Dobzhansky, Theodosius. 1955. "A Review of Some Fundamental Concepts and Problems in Population Genetics." *Cold Spring Harbor Symposium in Quantitative Biology* 20: 1–15.

———. 1962. *Mankind Evolving.* New Haven, CT: Yale University Press.

———. 1973. "Nothing in Biology Makes Sense Except in the Light of Evolution." *American Biology Teacher,* vol. 35: http://www.phil.vt.edu/Burian/NothingInBiolChFina.pdf.

Dobzhansky, Theodosius, and L. C. Dunn. 1952 [1946]. *Heredity, Race, and Society.* Rev. ed. New York: Pelican.

Doggett, Tom. 2002. "Crawling Snakehead Fish Scaring Washington," Reuters, July 28.

Dudley, Rachel. 2012. "Toward An Understanding of the 'Medical Plantation' as a Cultural Location of Disability." *Disability Studies Quarterly* 32 (4).

Duke Homestead Tour, 1990. Information from a personal tour of Duke Homestead, http://www.nchistoricsites.org/duke/.

Dumit, Joseph. 2012. "Prescription Maximization and the Accumulation of Surplus Health in the Pharmaceutical Industry: The BioMarx Experiment." In *Lively Capital: Biotechnologies, Ethics and the Governance in Global Markets,* edited by Kaushik Sunder Rajan. Durham, NC: Duke University Press.

Duster, Troy. 2003. *Backdoor to Eugenics.* 2nd ed. New York: Routledge.

Egan, Timothy. 2011. "Frankenfish Phobia," *The New York Times,* March 17, http://opinionator.blogs.nytimes.com/2011/03/17/frankenfish-phobia/.

Eisenhart, Margaret A., and Elizabeth Finkel. 1998. *Women's Science: Learning and Succeeding from the Margins.* Chicago: University of Chicago Press.

Ekberg, Merryn. 2007. "The Old Eugenics and the New Genetics Compared." *Social History of Medicine* 20 (3): 581–93.

Elmore, C. D., H. R. Hurst, and D. F. Austin. 1990. "Biology and Control of Morningglory (*Ipomoea* spp.)." *Journal Review of Weed Science* 5: 83–114.

Else-Quest, N. M., J. S. Hyde, and M. C. Linn. 2010. "Cross-national Patterns of Gender Differences in Mathematics: A Meta-Analysis." *Psychological Bulletin* 136: 103–27.

Elton, Charles C. 1958. *The Ecology of Invasion by Animals and Plants.* London: Methuen.

Emery, David. 2002. "Attack of the Frankenfish: Northern Snakehead Fish Invades North America," About.com Urban Legends, http://urbanlegends.about.com/od/fish/a/snakehead.htm.

Enserink, Martin. 1999. "Biological Invaders Sweep In." *Science* 285 (5435): 1834–36.

Epperson, Bryan K. 1986, December. "Spatial-Autocorrelation Analysis of Flower Color Polymorphisms within Sub-structured Populations of Morning Glory (*Ipomoea purpurea*)." *The American Naturalist* 128 (6): 840–58.

Epperson, Bryan K., and Michael T. Clegg. 1983. "Flower Color Variation in the Morning Glory, *Ipomoea purpurea.*" *Journal of Heredity* 74 (4): 247–50.

Eriksen, Thomas Hylland. 2006. "Diversity versus Difference: Neo-liberalism in the Minority Debate." In *The Making and Unmaking of Difference*, edited by Richard Rottenburg, Burkhard Schnepel, and Shingo Shimada, 13–36. Bielefeld, Ger.: Transaction.

Escobar, Arturo. 1997. "Cultural Politics and Biological Diversity: State, Capital, and Social Movements in the Pacific Coast of Colombia." In *The Politics of Culture in the Shadow of Capital,* edited by Lisa Lowe and David Lloyd. Durham, NC: Duke University Press.

Etzkowitz, Henry, Carol Kemelgor, Michael Neuschatz, and Brian Uzzi. 1994. "Barriers to Women in Academic Science and Engineering." In *Who Will do Science? Educating the Next Generation*, edited by Willie Pearson Jr. and Irwin Fechter. Baltimore: Johns Hopkins University Press.

Faulkner, William. 1951. *Requiem for a Nun.* New York: Random House.

Fausto-Sterling, Anne. 1985. *Myths of Gender: Biological Theories About Men and Women.* New York: Basic.

———. 1987. "Society Writes Biology/Biology Constructs Gender." *Daedalus* 116 (4): 61–76.

———. 1995. "Gender, Race, and Nation: The Comparative Anatomy of 'Hottentot' Women in Europe, 1815–1817." In *Deviant Bodies: Critical Perspectives on Difference in Science and Popular Culture,* edited by Jennifer Terry and Jacqueline Uria, 19–48. Bloomington: Indiana University Press.

———. 2003. "Science Matters, Culture Matters." *Perspectives in Biology and Medicine* 46 (1): 109–24.

Feldman, Daniel C., and William H. Turnley. 2004. "Contingent Employment in Academic Careers: Relative Deprivation among Adjunct Faculty." *Journal of Vocational Behavior* 64 (2): 284–307.

Fields, Helen. 2005, February. "Invasion of the Snakeheads." *Smithsonian Magazine.*

Fineblum, Wendy L., and Mark D. Rausher. 2002, October 12. "Tradeoff between Resistance and Tolerance to Herbivore Damage in a Morning Glory." *Nature* 377: 517–20.

Fisher, Jill. 2011. *Gender and the Science of Difference: Cultural Politics of Contemporary Science and Medicine.* New Brunswick, NJ: Rutgers University Press.

Fisher, R. A. 1941. "Average Excess and Average Effect of Gene Substitution." *Annals of Eugenics* 11: 53–63.

Forrest, G. I., and A. F. Fletcher. 1995. "Implication of Genetic Research for Pinewood

Conservation." In *Our Pinewood Heritage*, edited by J. R. Aldhous, 97–106. Farnham: Royal Society for the Protection of Birds.

Foucault, Michel. 1977. *Discipline and Punish: The Birth of the Prison*. New York: Pantheon.

———. 1980. *Power/Knowledge: Selected Interviews and Other Writings, 1972–1977*. New York: Pantheon.

Fox, Mary Frank. 2000. "Organizational Environments and Doctoral Degrees Awarded to Women in Science and Engineering Departments." *Women's Studies Quarterly* 28: 47–61.

Fox, Mary Frank, Gerhard Sonnert, and Irina Nikiforova. 2011, October. "Programs for Undergraduate Women in Science and Engineering: Issues, Problems, and Solutions." *Gender & Society* 22 (5): 589–615.

Frazier, Ian. 2010, October 25. "Fish Out of Water: The Asian Carp Invasion." *New Yorker*, 66–73.

Freedman, Estelle. 2002. *No Turning Back: The History of Feminism and the Future of Women*. New York: Ballantine.

Fuller, Laurie, and Erica R. Meiners. 2005. "Reflections: Empowering Women, Technology, and (Feminist) Institutional Changes." *Frontiers: A Journal of Women Studies* 26 (1): 168–80.

Fullwiley, D. 2007. "The Molecularization of Race: Institutionalizing Human Difference in Pharmacogenetics Practice." *Science as Culture* 16 (1): 1–30.

Galton, Francis. 1865. "Hereditary Talent and Character." *Macmillan's Magazine* 12: 163.

———. 1909. *Memories of My Life*. 3rd ed. London: Methuen.

Gannett, Lisa. 2001. "Racism and Human Genome Diversity Research: The Ethical Limits of 'Population Thinking.'" *Philosophy of Science* 68 (3): S479–S492.

Garforth, Lisa, and Anne Kerr. 2009, Fall. "Women and Science: What's the Problem?" *Social Politics: International Studies in Gender, State and Society* 16 (3): 379–403.

Gates, Barbara T., and Ann B. Shteir. 1997. *Natural Eloquence: Women Reinscribe Science*. Madison: University of Wisconsin Press.

Gere, Cathy. 1999. "Bones That Matter: Sex Determination in Paleodemography, 1948–1995." *Studies in the History and Philosophy of Biology and the Biomedical Sciences* 30 (4): 455–71.

Ghiselin, Michael. T. 1971. "The Individual in the Darwinian Revolution." *New Literary History* 3 (1): 113–34.

———. 1995, December. "Darwin, Progress, and Economic Principles." *Evolution* 49 (6): 1029–37.

———. 2005. "The Darwinian Revolution as Viewed by a Philosophical Biologist." *Journal of the History of Biology* 38 (1): 123–36.

Gigerenzer, Gerd, Zeno Swijtink, Theodore Porter, Lorraine Daston, John Beatty, and Lorenz Kruger. 1989. *The Empire of Chance—How Probability Changed Science and Everyday Life*. Cambridge: Cambridge University Press.

Gilbert, Scott, Jan Sapp, and Alfred Tauber. 2012. "A Symbiotic View of Life: We Have Never Been Individuals." *Quarterly Review of Biology* 87 (4): 325–41.

Gilbert, S. F. 2002. "Genetic Determinism: The Battle between Scientific Data and Social

Image in Contemporary Developmental Biology." In *On Human Nature: Anthropological, Biological, and Philosophical Foundations,* edited by A. Grunwald, M. Gutmann, and E. M. Neumann-Held, 121–40. New York: Springer-Verlag.

Gilham, Nicholas. 2001. *A Life of Sir Francis Galton: From African Exploration to the Birth of Eugenics.* Oxford: Oxford University Press.

Gilman, Sander. 1985. *Difference and Pathology: Stereotypes of Sexuality, Race, and Madness.* Ithaca, NY: Cornell University Press.

Glover, Deborah, Mary L. Durbin, Gavin Huttley, and Michael T. Clegg. 1996. "Genetic Diversity in the Common Morning Glory." *Plant Species Biology* 11: 41–50.

Gobster, P. H. 2005. "Invasive Species as Ecological Threat: Is Restoration an Alternative to Fear-based Resource Management?" *Ecological Restoration* 23: 261–70.

Goonatilake, Susantha. 1984. *Aborted Discovery: Science and Creativity in the Third World.* London: Zed.

Gonsalves, Allison J. 2011. "Gender and Doctoral Physics Education: Are We Asking the Right Question?" In *Doctoral Education: Research Based Strategies for Doctoral Students, Supervisors, and Administrators,* edited by L. McAlpine and C. Amundsen. Springer Science+Business Media B.V.

Gordon, Avery. 1997. *Ghostly Matters: Haunting and the Sociological Imagination.* Minneapolis: University of Minnesota Press.

Gould, Stephen Jay. 1979, December. "Darwin's Middle Road." *Natural History* 88 (10): 27–31.

———. 1980. "Wallace's Fatal Flaw." *Natural History* 89 (1): 26–40.

———. 1981. *The Mismeasure of Man.* New York: Norton.

———. 1982. *The Structure of Evolutionary Theory.* Cambridge, MA: Harvard University Press.

———. 1991, December. "The Smoking Gun of Eugenics." *Natural History* 100 (12), http://homepages.dcc.ufmg.br/~assuncao/pgm/aulas/TheSmokingGunOfEugenics.pdf.

———. 1997. "An Evolutionary Perspective on Strengths, Fallacies, and Confusions in the Concept of Native Species." In *Nature and Ideology: Natural Garden Design in the Twentieth Century,* edited by Joachim Wolschke-Bulmahn, vol. 18: 11–19. Washington, DC: Dumbarton Oaks Research Library and Collection.

———. 2002. "An Evolutionary Perspective on the Concept of Native Species." In *I Have Landed: The End of a Beginning in Natural History,* 335–46. New York: Harmony.

Greco, M. 1993. "Psychosomatic Subjects and the 'Duty to Be Well': Personal Agency within Medical Rationality." *Economy and Society* 22 (3): 357–72.

Gregory, Richard. 1998. "Brainy Mind." *British Medical Journal* 317: 1693–95.

Grene, Marjorie. 1990, July. "Evolution, 'Typology' and 'Population Thinking.'" *American Philosophical Quarterly* 27 (3): 237–44.

———, ed. 1983. *Dimensions of Darwinism: Themes and Counterthemes in Twentieth-Century Evolutionary Theory.* Cambridge: Cambridge University Press.

Grese, Robert E. 2011. "Introduction." In *The Native Landscape Reader,* 3–22. Amherst: University of Massachusetts Press.

Gromley, Melinda. 2009. "Scientific Discrimination and the Activist Scientist: L. C. Dunn and the Professionalization of Genetics and Human Genetics in the United States." *Journal of the History of Biology* 42: 33–72.

Groning, G., and J. Wolschke-Bulmahn. 2003. "The Native Plant Enthusiasm: Ecological Panacea or Xenophobia?" *Landscape Research* 28: 75–88.

Grosz, Elizabeth. 1994. *Volatile Bodies: Toward a Corporeal Feminism*. Sydney: Allen and Unwin.

———. 2011. *Becoming Undone: Darwinian Reflections on Life Politics and Art*. Durham, NC: Duke University Press.

Gudding, Gabriel. 1996. "The Phenotype/Genotype Distinction and the Disappearance of the Body." *Journal of the History of Ideas* 57 (3): 525–45.

Haack, Robert A., Frank Herard, Jianghua Sun, and Jean J. Turgeon. 2010. "Managing Invasive Populations of Asian Longhorned Beetle and Citrus Longhorned Beetle: A Worldwide Perspective." *Annual Review of Entomology* 55: 521–46.

Hacker, Andrew, and Claudia Dreyfus. 2011. *Higher Education: How Colleges are Wasting Our Money and Failing Our Kids—and What We Can Do about It*. New York: St. Martin's Griffin.

Hall, Roberta M., and Bernice R. Sandler. 1982. *The Classroom Climate: A Chilly One for Women?* Washington, DC: Association of American Colleges.

———. 1984. *Out of the Classroom: A Chilly Climate for Women?* Washington DC: Association of American Colleges.

Haller, Mark. 1984. *Eugenics: Hereditarian Attitudes in American Thought*. New Brunswick, NJ: Rutgers University Press.

Hammonds, Evelynn. 1993. "Never Meant to Survive: A Black Woman's Journey." In *The "Racial" Economy of Science: Towards a Democratic Future*, edited by Sandra Harding, 239–48. Bloomington: Indiana University Press.

———. 1994, Summer/Fall. "Black (W)holes and the Geometry of Black Female Sexuality." *differences: A Journal of Feminist Cultural Studies* 6 (2/3): 126–46.

———. 1999. "The Logic of Difference: A History of Race in Science and Medicine in the United States." Presentation at the Women's Studies Program, UCLA.

Hammonds, Evelynn, and Rebecca M. Herzig. 2009. *The Nature of Difference: Sciences of Race in the United States from Jefferson to Genomics*. Cambridge, MA: MIT Press.

Hancock, Ange-Marie, and Nira Yuval-Davis, eds. 2011. *The Politics of Intersectionality*. Palgrave-Macmillan.

Handelsman, Jo, et al. 2005. "More Women in Science." *Science* 309 (5738): 1190–91.

Hansen, Nancy, Heidi L Janz, and Dick J. Sobsey. 2008. "21st Century Eugenics." *The Lancet* 372: 104–7.

Hanson, Sandra L. 2004, Spring. "African American Women in Science: Experiences from High School through the Post-Secondary Years and Beyond." *Feminist Formations* 16 (1): 96–115.

Haraway, Donna J. 1988. "Situated Knowledges: The Science Question in Feminism and the Privilege of Partial Perspective." *Feminist Studies* 14 (3): 575–99.

———. 1989. *Primate Visions: Gender, Race, and Nature in the World of Modern Science.* New York: Routledge.

———. 1991. *Simians, Cyborgs, and Women: The Reinvention of Nature.* New York: Routledge.

———. 1997. *Modest_Witness@Second_Millennium: FemaleMan__Meets_Onco Mouse.* New York: Routledge.

———. 1999. *How Like a Leaf: An Interview with Thyrza Nichols Goodve.* New York: Routledge.

———. 2007. *When Species Meet.* Minneapolis: University of Minnesota Press.

Harding, Sandra. 1986. *The Science Question in Feminism.* Milton Keynes, UK: Open University Press.

———. 1991. *Whose Science? Whose Knowledge?* Milton Keynes, UK: Open University Press.

———. 1993. *The "Racial" Economy of Science: Toward a Democratic Future.* Bloomington: Indiana University Press.

———. 1998. *Is Science Multicultural? Postcolonialisms, Feminisms, and Epistemologies.* Bloomington: Indiana University Press.

———. 2006. *Science and Social Inequality: Feminist and Postcolonial Issues.* Urbana: University of Illinois Press.

———. 2008. *Sciences From Below: Feminisms, Postcolonialities, and Modernities.* Durham, NC: Duke University Press.

———. 2011. *The Postcolonial Science and Technology Studies Reader.* Durham, NC: Duke University Press.

Harley, J. L., and S. E. Smith. 1983. *Mycorrhizal Symbiosis.* London: Academic.

Harmon, Amy. 2007. "Prenatal Test Puts Down Syndrome in Hard Focus," *The New York Times,* May 9.

Harrison, Carlos. 2002. "Something from the X-Files," *Miami New Times,* August 29.

Harrison, Faye V. 1996. "Anthropology, Fiction, and Unequal Relations of Intellectual Production." In *Women Writing Culture,* edited by Ruth Behar and Deborah Gordon, 233–45. Berkeley: University of California Press.

Hart, Roger. 1999. "On the Problem of Chinese Science." In *The Science Studies Reader,* edited by Mario Biagioli. New York: Routledge.

Hartmann, Betsy. 2010. "Rethinking Climate Refugees and Climate Conflict: Rhetoric, Reality and the Politics of Policy Discourse." *Journal of International Development* 22: 233–46.

———. 2011, Spring. "The Return of Population Control: Incentives, Targets and the Backlash against Cairo." *Different Takes,* no. 70, http://popdev.hampshire.edu/sites/default/files/uploads/u4763/DT%2070%20Hartmann.pdf.

Hartmann, Betsy, Banu Subramaniam, and Charles Zerner. 2005. *Making Threats: Biofears and Environmental Anxieties.* Lanham, MD: Rowman & Littlefield.

Hartmann, Elizabeth. 1995. *Reproductive Rights and Wrongs: The Global Politics of Population Control.* Brooklyn: South End.

Harwood, Jonathan. 1989. "Genetics, Eugenics and Evolution." *British Journal for the History of Science* 22 (3): 257–65.

Hattingh, Johan. 2011. "Conceptual Clarity, Scientific Rigour and 'the Stories We Are': Engaging with Two Challenges to the Objectivity of Invasion Biology." In *Fifty Years of*

Invasion Ecology: The Legacy of Charles Elton, edited by David M. Richardson, 359–75. Oxford: Blackwell.

Head, Lesley, and Jennifer Atchison. 2009. "Cultural Ecology: Emerging Human-Plant Geographies." *Progress in Human Geography* 33 (2): 236–45.

Hearn, Jeff. 2001. "Academia, Management, and Men: Making the Connections, Exploring the Implications." In *Gender and the Restructured University: Changing Management and Culture in Higher Education,* edited by Anne Brooks and Alison Mackinnon, 69–89. New York: Open University Press.

Hebert, Josef. 1998. "Feds to Fight Invaders." *ABC News,* February 3.

Helmreich, Stefan. 2009. *Alien Ocean: Anthropological Voyages in Microbial Seas.* Berkeley: University of California Press.

Hemmings, Clare. 2011. *Why Stories Matter: The Political Grammar of Feminist Theory.* Durham, NC: Duke University Press.

Henrion, Claudia. 1997. *Women in Mathematics: The Addition of Difference.* Bloomington: Indiana University Press.

Henson, Pamela M. 1987. "The Comstocks of Cornell: A Marriage of Interests." In *Creative Couples in the Sciences,* edited by Helena Pycior, Nancy Slack, and Pnina Abir-Am, 112–25. New Brunswick, NJ: Rutgers University Press.

Hess, David. 1997. *Science Studies: An Advanced Introduction.* New York: NYU Press.

Hettinger, Ned. 2011, May. "Exotic Species, Naturalization, and Biological Nativism." *Environmental Values* 10 (2): 193–224.

Hierro, Jose, Diego Villare, Ozkan Eren, Jon M. Graham, and Ragan M. Callaway. 2006, August. "Disturbance Facilitates Invasion: The Effects Are Stronger Abroad than at Home." *The American Naturalist* 168 (2): 145–56.

Hill, Constance, Christianne Corbett, and Andresse St. Rose. 2010. "Why So Few? Women in Science, Technology, Engineering and Mathematics," American Association of University Women, http://www.aauw.org/learn/research/upload/whysofew.pdf.

Hird, M. J. 2004. *Sex, Gender, and Science.* London: Palgrave.

Hobbs, Richard J., and David M. Richardson. 2011. "Invasion Ecology and Restoration Ecology: Parallel Evolution in Two Fields of Endeavour." In *Fifty Years of Invasion Ecology: The Legacy of Charles Elton,* edited by David M. Richardson, 61–70. Oxford: Blackwell.

Hodges. Sarah. 2008. *Contraception, Colonialism and Commerce.* Farnham, UK: Ashgate.

———. 2010. "South Asia's Eugenic Pasts." In *The Oxford Handbook of the History of Eugenics,* edited by Alison Bashford and Philippa Levine, 228–42. New York: Oxford University Press.

Hodgson, Dennis. 1991. "The Ideological Origins of the Population Association of America." *Population and Development Review* 17 (1): 1–34.

Holden, Constance. 2000. "Parity as a Goal Sparks Bitter Battle." *Science* 289 (5478): 380.

Hubbard, Ruth. 1983. "Have Only Men Evolved?" In *Discovering Reality: Feminist Perspectives on Epistemology, Metaphysics, Methodology, and Philosophy of Science,* edited by Sandra Harding and Merrill B. Hintikka, 45–69. Dordrecht, Neth.: Kluwer Academic.

———. 1990. *The Politics of Women's Biology.* New Brunswick, NJ: Rutgers University Press.

Hubbard, Ruth, and Elijah Wald. 1999. *Exploding the Gene Myth: How Genetic Information Is Produced and Manipulated by Scientists, Physicians, Employers, Insurance Companies, Educators, and Law Enforcers*. Boston: Beacon.

Hulme, Philip E. 2011. "Biosecurity: The Changing Face of Invasion Biology." In *Fifty Years of Invasion Ecology: The Legacy of Charles Elton*, edited by David M. Richardson, 301–14. Oxford: Blackwell.

Hurston, Zora Neale. n.d. "Letter to Countee Cullen." PBS American Masters, http://www.pbs.org/wnet/americanmasters/episodes/zora-neale-hurston/introduction/93/.

Huslin, Anita. 2002a. "Biologists on Mission to Kill," *Washington Post,* August 18, C03.

———. 2002b. "Freaky Fish Story Flourishes," *Washington Post,* July 3, B01.

———. 2002c. "U. S. Moves to Ban Import of Snakeheads," *Washington Post,* July 23, B03.

Huslin, Anita, and Michael Ruane. 2002. "Spawn of Snakehead?" *Washington Post,* July 10, B01.

Hyde, J. S., S. M. Lindbert, M. C. Linn, A. B. Ellis, and C. C. Williams. 2008. "Gender Similarities Characterize Math Performance." *Science* 321: 494–95.

Hynes, Patricia H. 1989. *The Recurring Silent Spring*. New York: Pergamon.

Intemann, Kristen. 2009. "Why Diversity Matters: Understanding and Applying the Diversity Component of the National Science Foundation's Broader Impacts Criterion." *Social Epistemology* 23 (3/4): 249–66.

Iverson, Susan VanDeventer. 2007. "Camouflaging Power and Privilege: A Critical Race Analysis of University Diversity Policies." *Educational Administration Quarterly* 43: 586–611.

Jacob, Margaret C. 1998. "Latour's Version of the Seventeenth Century." In *A House Built on Sand: Exposing Postmodernist Myths About Science,* edited by Noretta Koertge, 240–54. New York: Oxford University Press.

Jamieson, D. 1995. "Ecosystem Health: Some Preventive Medicine." *Environmental Values* 4: 333–44.

Jansson, Stefan, and Carl J. Douglas. 2007. "Populus: A Model System for Plant Biology." *Annual Review of Plant Biology* 58: 435–58.

Johnson, Dawn R. 2011, Winter. "Women of Color in Science, Technology, Engineering, and Mathematics (STEM)." *New Directions for Institutional Research* 2011 (152): 75–85.

Kahn, Jonathan. 2005, November/December. "BiDil: False Promises." *Gene Watch* 18 (6): 6–9.

Kaiser, Jocelyn. 1999. "Stemming the Tide of Invading Species." *Science* 285 (5435): 1836–40.

Kamin, Leon J. 1974. *The Science and Politics of IQ*. New York: Halsted.

Kaplan, Kim. 1991, July. "USDA Plant Hunters: Bring 'Em Back Alive and Growing." *Agricultural Research,* 4–12, http://pubpages.unh.edu/~gec/IA401/USDA%20Plant%20Hunters.PDF.

Kass-Simon, G., and Patricia Farnes. 1990. *Women of Science: Righting the Record*. Bloomington: Indiana University Press.

Kaufman, Sharon R., and Lynn M. Morgan. 2005. "The Anthropology of the Beginnings and Ends of Life." *Annual Review of Anthropology* 34: 317–41.

Keller, David R., and Frank B. Golley. 2000. *The Philosophy of Ecology: From Science to Synthesis*. Athens: University of Georgia Press.

Keller, Evelyn Fox. 1977. "An Anomaly of a Woman in Physics." In *Working It Out: 23 Women Writers, Artists, Scientists, and Scholars Talk About Their Lives and Work*, edited by Sara Ruddick and Pamela Daniels. New York: Pantheon.

———. 1983. *A Feeling for the Organism: The Life and Work of Barbara McClintock*. San Francisco: Freeman.

———. 1985. *Reflections on Gender and Science*. New Haven, CT: Yale University Press.

———. 1987. "Is Gender/Science System: or, Is Sex to Gender as Nature is to Science?" *Hypatia* 2 (3): 37–49.

———. 1992. "How Gender Matters, Or, Why It's So Hard for Us to Count Past Two." In *Inventing Women: Science, Technology and Gender*, edited by Gill Kirkup and Laurie Smith Keller, 42–56. Cambridge: Polity, in association with the Open University.

———. 1996. "The Biological Gaze." In *Future Natural: Nature/Science/Cultural*, edited by G. Roberson, M. Mash, L. Tickner, J. Bird, B. Curtis, and T. Putnam, 107–21. London: Routledge.

———. 2001. "Making a Difference: Feminist Movement and Feminist Critiques of Science." In *Feminism in Twentieth-Century Science, Technology, and Medicine*, edited by Andrea N. H. Creager, Elizabeth Lunbeck, and Londa Schiebinger. Chicago: University of Chicago Press.

———. 2010. *The Mirage of a Space between Nature and Nurture*. Durham, NC: Duke University Press.

Keller, Evelyn Fox, and Helen Longino, eds. 1996. "Introduction." In *Feminism and Science*, 1–14. Oxford: Oxford University Press.

Kerr, Anne. 1998. "Eugenics and the New Genetics in Britain: Examining Contemporary Professionals' Accounts." *Science, Technology, & Human Values* 23 (2): 175–98.

Kevles, Daniel, J. 1985. *In the Name of Eugenics: Genetics and the Uses of Human Heredity*. New York: Knopf.

———. 1998. *In the Name of Eugenics: Genetics and the Uses of Human Heredity*. Cambridge, MA: Harvard University Press.

Keynes, John Maynard. 1926. *The End of Laissez-faire*, http://www.panarchy.org/keynes/laissezfaire.1926.html.

Khalil, Elias. 2000. "Beyond Natural Selection and Divine Intervention: The Lamarckian Implication of Adam Smith's Invisible Hand." *Journal of Evolutionary Economics* 10: 373–93.

Kim, Kyung-Man. 1994. *Explaining Scientific Consensus: The Case of Mendelian Genetics*. New York: Guilford.

Kirby, V. 1997. *Telling Flesh: The Substance of the Corporeal*. New York: Routledge.

Kirksey, S. Eben, and Stefan Helmreich. 2010. "The Emergence of Multispecies Ethnography." *Cultural Anthropology* 25 (4).

Kitcher, Philip. 1997. *The Lives to Come: The Genetic Revolution and Human Possibilities*. New York: Simon and Schuster.

Klausen, Susanne, and Alison Bashford, 2010. "Eugenics, Feminism and Fertility Control." In *The Oxford Handbook of the History of Eugenics*, edited by Alison Bashford and Philippa Levine. Oxford: Oxford University Press.

Klein, H. 2002. *Weeds, Alien Plants and Invasive Plants*. PPRI Leaflet Series: Weeds Biocontrol, no. 1.1. Pretoria: ARC-Plant Protection Research Institute.

Kline, Wendy. 2005. *Building a Better Race: Gender, Sexuality, and Eugenics from the Turn of the Century to the Baby Boom*. Berkeley: University of California Press.

Klinkenborg, Verlyn. 2013. "Hey, You Calling Me an Invasive Species?" *The New York Times*, September 8.

Kluchin, Rebecca. 2009. *Fit to Be Tied: Sterilization and Reproductive Rights in America, 1950–1980*. New Brunswick, NJ: Rutgers University Press.

Kluger, Jeffrey. 2002, July 29. "Fish Tale." *Time Magazine*, http://www.time.com/time/magazine/article/0,9171,332042,00.html.

Knorr Cetina, Karin. 1999. *Epistemic Cultures: How the Sciences Make Knowledge*. Cambridge, MA: Harvard University Press.

Kobell, Rona, and Candus Thomson. 2002. "State Shares Victory over Snakehead Fish," *Baltimore Sun*, November 21.

Kobert, Elizabeth. 2011, January 31. "America's Top Parent: What's Behind the 'Tiger Mother' Craze?" *The New Yorker*, 70–73.

Koblitz, Anne Hibner. 1983. *A Convergence of Lives: Sofia Kovalevskaia, Scientist, Writer, Revolutionary*. New Brunswick, NJ: Rutgers University Press.

Kohlstedt, Sally Gregory. 2004. "Sustaining Gains: Reflections on Women in Science and Technology in 20th Century United States." *Feminist Formations* 16 (1): 1–26.

———, ed. 1999. *History of Women in the Sciences: Readings from Isis*. Chicago: University of Chicago Press.

Kolata, Gina. 2011. "Women Atop Their Fields Dissect the Scientific Life," *The New York Times*, June 6.

Kreitman, Martin. 2000. "Methods to Detect Selection in Populations with Applications to the Human." *Annual Review of Genomics and Human Genetics* 1: 539–59.

Kumar, Neelam, ed. 2009. *Women and Science in India: A Reader*. New Delhi: Oxford University Press.

Kyung-Man, Kim. 1994. *Explaining Scientific Consensus: The Case of Mendelian Genetics*. New York: Guilford.

Larson, Brendon. 2005. "The War of the Roses: Demilitarizing Invasion Biology." *Frontiers in Ecology and the Environment* 3: 495–500.

———. 2007a. "An Alien Approach to Invasive Species: Objectivity and Society in Invasion Biology." *Biological Invasions* 9: 947–56.

———. 2007b. "Entangled Biological, Cultural and Linguistic Origins of the War on Invasive Species." In *Body, Language and Mind*, vol. 2, *Sociocultural Situatedness*, edited by R. Frank, R. Dirven, J. Zlatev, and T. Ziemke. New York: Mouton de Gruyter.

———. 2007c. "Thirteen Ways of Looking at Invasive Species." In *Invasive Plants: Inven-*

tories, Strategies and Action. Topics in Canadian Weed Science, vol. 5, edited by D. R. Clements and S. J. Darbyshire, 131–46. Quebec: Canadian Weed Science Society.

Laslett, Barbara, and Barrie Thorne, eds. 1997. *Feminist Sociology: Life Histories of a Movement.* New Brunswick, NJ: Rutgers University Press.

Latour, Bruno. 1993a, Winter. "Pasteur on Lactic Acid Yeast: A Partial Semiotic Analysis." *Configurations* 1: 129–46.

———. 1993b. *We Have Never Been Modern.* Cambridge, MA: Harvard University Press.

Latour, Bruno, and Steven Woolgar. 1986. *Laboratory Life: The Construction of Scientific Facts.* Princeton, NJ: Princeton University Press.

Lawrence, Peter A. 2006. "Men, Women, and Ghosts in Science." *PLoS Biology* 4 (1): 13–15.

Lear, Linda. 1997. *Rachel Carson: Witness for Nature.* Boston: Mariner.

Lederman, Muriel, and Ingrid Bartsch, eds. 2001. *The Gender and Science Reader.* New York: Routledge.

Lemke, Thomas. 2007. "Susceptible Individuals and Risky Rights: Dimensions of Genetic Responsibility." In *Biomedicine as Culture: Instrumental Practices, Technoscientific Knowledge, and New Modes of Life.* London: Routledge.

Leonard, Eileen B. 2003. *Women, Technology, and the Myth of Progress.* Upper Saddle River, NJ: Prentice Hall.

Lerdau, Manuel, and Jacob D. Wickham. 2011. "Non-Natives: Four Risk Factors." *Nature* 475: 36–37.

Levine, Philippa, and Alison Bashford. 2010. "Introduction: Eugenics and the Modern World." In *The Oxford Handbook of the History of Eugenics,* edited by Alison Bashford and Philippa Levine, 3–26. Oxford: Oxford University Press.

Levine, Susan. 1995. *Degrees of Equality: The American Association of University Women and the Challenge of Twentieth-Century Feminism.* Philadelphia: Temple University Press.

Levy, David, and Sandra Peart. 2004. "Statistical Prejudice: From Eugenics to Immigrants." *European Journal of Political Economy* 20: 5–22.

Lewontin, Richard. 1970. "The Units of Selection." *Annual Review of Ecology and Systematics* 1:1–18.

———. 1974. *The Genetic Basis of Evolutionary Change.* New York: Columbia University Press.

———. 1978. "Adaptation." *Scientific American* 239: 212–30.

———. 1987. "Polymorphism and Heterosis: Old Wine in New Bottles and Vice Versa." *Journal of the History of Biology* 20 (3): 337–49.

Lippincott, Gail. 2003. "Rhetorical Chemistry: Negotiating Gendered Audiences in Nineteenth-Century Nutrition Studies." *Journal of Business and Technical Communication* 17: 10–49.

Lockwood, Julie L., Martha F. Hoopes, and Michael P. Marchetti. 2011. "'Non Natives' Plusses of Invasion Ecology." *Nature* 475: 36.

Lohmann, Larry. 2005. "Malthusianism and the Terror of Scarcity." In *Making Threats: Biofears and Environmental Anxieties,* edited by Betsy Hartmann et al., 81–98. New York: Rowman & Littlefield.

Lombardo, Paul. 2002. "The American Breed: Nazi Eugenics and the Origins of the Pioneer Fund." *Albany Law Review* 65 (3): 743–830.

Long, Scott. J., ed. 2001. *From Scarcity to Visibility: Gender Differences in the Careers of Doctoral Scientists and Engineers.* Washington, DC: National Academy of Sciences.

Longino, Helen E. 1989. "Can There Be a Feminist Science?" In *Feminism and Science,* edited by Nancy Tuana. Bloomington: Indiana University Press.

———. 1990. *Science as Social Knowledge: Values and Objectivity in Scientific Inquiry.* Princeton, NJ: Princeton University Press.

Lorber, Judith. 2000. "Using Gender to Undo Gender: A Feminist Degendering Movement." *Feminist Theory* 1 (1): 79–95.

Lorde, Audre. 1984. "Age, Race, Class and Sex: Women Redefining Difference." In *Sister Outsider,* edited by Audre Lorde. Freedom, CA: The Crossing Press.

Louçã, Francisco. 2008. "Emancipation through Interaction—How Eugenics and Statistics Converged and Diverged." *Journal of the History of Biology* 42 (4): 649–84.

Lovett, Laura. 2007. *Conceiving the Future: Pronatalism, Reproduction, and the Family in the United States, 1890–1930.* Gender and American Culture Series. Chapel Hill: University of North Carolina Press.

Lovitts, Barbara. 2007. *Making the Implicit Explicit: Creating Performance Expectations for the Dissertation.* Sterling, VA: Stylus.

Mabberley, D. J. 1997. *The Plant-Book: A Portable Dictionary of the Vascular Plants.* 2nd ed. Cambridge: Cambridge University Press.

Mabey, R. 2005. "From Corn Poppies to Eagle Owls." *ECOS* 26 (3/4): 41–46.

MacAusland, Carol, and Christopher Costello. 2004, September. "Avoiding Invasives: Trade Related Policies for Controlling Unintentional Exotic Species Introductions." *Journal of Environmental Economics and Management* 45 (2): 954–77.

Macdougall, A. S., and R. Turkington. 2005. "Are Invasive Species the Drivers or Passengers of Change in Degraded Ecosystems?" *Ecology* 86: 42–55.

Mack, Richard N., et al. 2000, Spring. "Biotic Invasions: Causes, Epidemiology, Global Consequence and Control." *Issues in Ecology* 15 (5): 12.

MacKenzie, Donald A. 1981. *Statistics in Britain, 1865–1930: The Social Construction of Scientific Knowledge.* Edinburgh: Edinburgh University Press.

Maddox, Brenda. 2002. *Rosalind Franklin: The Dark Lady of DNA.* New York: HarperCollins.

Magubane, Zine. 2012. "Which Bodies Matter? Feminism, Poststructuralism, Race and the Curious Theoretical Odyssey of the 'Hottentot Venus.'" *Gender and Society* 26 (6): 816–34.

Malcolm, Shirley M. 1999. "Fault Lines." *Science* 284: 1271.

Malone, Scott. 2012. "U.S. Economy Losing Competitive Edge: Survey," Reuters, January 18, http://www.reuters.com/article/2012/01/18/us-corporate-competitiveness-idUSTRE80H1HR20120118.

Mann, Charles C. 2011, May/June. "The Dawn of the Homogenocene: Tracking Globalization Back to Its Roots." *Orion Magazine.*

Mansfield, Harvey. 2006. *Manliness*. New Haven, CT: Yale University Press.

Marais, Robert de. 1974. "The Double-Edged Effect of Sir Francis Galton: A Search for the Motives in the Biometrician-Mendelian Debate." *Journal of the History of Biology* 7: 141–74.

Margulis, Lynn. 1998. *Symbiotic Planet: A Look at Evolution*. Amherst, MA: Perseus.

Margulis, Lynn, and Dorion Sagan. 2002. *Acquiring Genomes: A Theory of the Origins of Species*. New York: Basic.

Marinelli, Janet, and John M. Randall. 1996. *Invasive Plants: Weeds of the Global Garden*. Brooklyn: Brooklyn Botanic Garden.

Marklein, Mary Beth. 2012. "Record Number of Foreign Students in U.S.," *USA Today*, November 12.

Markowitz, Sally. 2001. "Pelvic Politics: Sexual Dimorphism and Racial Difference." *Signs* 26 (2): 389–414.

Marler, Marilyn, Catherine Zabinksi, and Ragan Callaway. 1999, June. "Mycorrhizae Indirectly Enhance Competitive Effects of an Invasive Forb on a Native Bunchgrass." *Ecology* 80 (4): 1180–86.

Martin, Emily. 1992. *The Woman in the Body*. Boston: Beacon.

———. 1997. "The End of the Body?" In *The Gender Sexuality Reader*, edited by Roger N. Lancaster and Micaela di Leonardo, 543–59. New York: Routledge.

———. 2006. "The Pharmaceutical Person." *BioSocieties* 1: 273–87.

Martinez, Elisabeth D., et al. 2007. "Falling Off the Academic Bandwagon." *European Molecular Biology Organization Reports* 8 (11): 977–81.

Marvier, Michelle, Peter Kareiva, and Michael Neubert. 2004. "Habitat Destruction, Fragmentation, and Disturbance Promote Invasion by Habitat Generalists in a Multispecies Metapopulation." *Risk Analysis* 24 (4): 869–78.

"Maryland Fears Carnivore Fish Invasion." *Edmonton Journal*. 2002. July 13.

Mason, M. A., and M. Goulden. 2004, November/December. "Do Babies Matter Part II? Closing the Baby Gap." *Academe*: 11–15.

Matt, Susan J. 2012. "The New Globalist is Homesick," *The New York Times*, March 21.

Matthews, Christine M. 2010. "Foreign Science and Engineering Presence in U.S Institutions and the Labor Force." Congressional Research Service 97–746, http://www.fas.org/sgp/crs/misc/97-746.pdf.

Mayberry, Maralee, Banu Subramaniam, and Lisa Weasel, eds. 2001. *Feminist Science Studies: A New Generation*. New York: Routledge.

Mayell, Hillary. 2002, July 2. "Maryland Wages War on Invasive Walking Fish." *National Geographic*.

Mayr, Ernst. 1973. "The Recent Historiography of Genetics." *Journal of the History of Biology* 6 (1): 125–54.

———. 1976. *Evolution and the Diversity of Life*. Cambridge, MA: Harvard University Press.

———. 1982. *The Growth of Biological Thought*. Cambridge, MA: The Belknap Press of Harvard University Press.

Mazumdar, Pauline M. 1991. *Eugenics, Human Genetics and Human Failings: The Eugenics Society, its Sources and its Critics in Britain*. London: Routledge.

———. 2002. "'Reform' Eugenics and the Decline of Mendelism." *Trends in Genetics* 18 (1): 48–52.

McCall, Leslie. 2005, Spring. "The Complexity of Intersectionality." *Signs* 30 (3).

McCann, Carole R. 2009. "Malthusian Men and Demographic Transitions." *Frontiers* 30 (1): 142–72.

McDonald, Andrew. 1991. "Origin and Diversity of Mexican Convolvulaceae." *Anales del Instituto de Biologia. Serie Botanica* 62 (1): 65–82.

McDonald, Kim A. 1999. "Biological Invaders Threaten U.S. Ecology." *Chronicle of Higher Education* 45 (23): A15.

McIntosh, Peggy. 2003. "White Privilege: Unpacking the Invisible Knapsack." In *Understanding Prejudice and Discrimination*, edited by Scott Plous, 191–96. New York: McGraw Hill.

McNeely, J. A. 2001. *The Great Reshuffling: Human Dimensions of Invasive Alien Species.* Gland, Switzerland: IUCN.

Mellström, Ulf. 2009. "The Intersection of Gender, Race and Cultural Boundaries, or Why is Computer Science in Malaysia Dominated by Women?" *Social Studies of Science* 39 (6): 885–907.

Merchant, Carolyn. 1980. *The Death of Nature: Women, Ecology, and the Scientific Revolution.* San Francisco: Harper & Row.

Merton, R. 1968. "The Matthew Effect in Science." *Science* 159 (3810): 56–63.

Mervis, Jeffrey. 2000. "Diversity: Easier Said Than Done." *Science* 289 (5478): 378–79.

Meyerson, Debra E., and Deborah M. Kolb. 2000. "Moving out of the Armchair: Developing a Framework to Bridge the Gap Between Feminist Theory and Practice." *Organization* 7(4): 553–72.

Mills, E. L. , J. H. Leach, J. T. Carlton, and C. L. Secor. 1994. "Exotic Species and the Integrity of the Great Lakes: Lessons from the Past." *Bioscience* 33 (10): 666–76.

Mills, E. L., M. D. Scheurell, D. L. Strayer, and J. T. Carlton. 1996. "Exotic Species in the Hudson River Basin: A History of Invasions and Introductions." *Estuaries* 19 (4): 814–23.

MIT (Massachusetts Institute of Technology). 1999. *A Study of the Status of Women Faculty in Science at MIT,* Special Issue of the MIT Faculty Newsletter 9 (March), http://web.mit.edu/fnl/women/women/html.

Moallem, Minoo, and Iain A. Boal. 1999. "Multicultural Nationalism and the Poetics of Inauguration." In *Between Women and Nation*, edited by Caren Kaplan, Norma Alarcon, and Minoo Moallem, 243–63. Durham, NC: Duke University Press.

Mohanty, Chandra. 2003. *Feminism Without Borders: Decolonizing Theory, Practicing Solidarity.* Durham, NC: Duke University Press.

Moi, Barbara. 2009. "A Dawn Tea Ceremony for the First Blooming of the Morning Glory." *Moebius* 7 (1): 31–38.

Moore, James. R. A. 2007. "Fisher: A Faith Fit for Eugenics." *Studies in the History and Philosophy of Science* 38 (1): 110–35.

Morland, Iain. 2011. "Intersex Treatment and the Promise of Trauma." In *Gender and the*

Science of Difference: Cultural Politics of Contemporary Science and Medicine, edited by Jill Fisher, 147–63. New Brunswick, NJ: Rutgers University Press.

Morley, B. D., and H. R. Toelken. 1983. *Flowering Plants in Australia.* Adelaide, Australia: Rigby.

Morley, Louise. 1999. *Organizing Feminisms: The Micropolitics of the Academy.* London: Macmillan.

———. 2003. *Quality and Power in Higher Education.* Berkshire, UK: Society for Research into Higher Education/Open University Press.

———. 2006. "Hidden Transcripts: The Micropolitics of Gender in Commonwealth Universities." *Women's Studies International Forum* 29: 543–51.

Morrison, Toni. 1992. *Playing in the Dark: Whiteness and the Literary Imagination.* New York: Vintage.

Moss-Racusin, Corinee A., John F. Dovidio, Victoria L. Brescoll, Mark J. Graham, and Jo Handelsman. 2012, September 17. "Science Faculty's Subtle Gender Biases Favor Male Students." *Proceedings of the National Academy of Science.* DOI 10:1073/pnas.1211286109

Mottier, Véronique. 2008. "Eugenics, Politics and the State: Social Democracy and the Swiss 'Gardening State.'" *Studies in History and Philosophy of Biology and Biomedical Sciences* 39: 263–69.

Muller, H. J. 1984. *Out of the Night: A Biologist's View of the Future.* New York: Garland.

Murray, T. R., D. A. Frank, and C. A. Gehring. 2010, March. "Ungulate and Topographic Control of Arbuscular Mychorrhizal Fungal Spore Community Composition in a Temperate Grassland." *Ecology* 91 (3): 815–27.

Nandy, Ashis. 1995. *Alternative Sciences: Creativity and Authenticity in Two Indian Scientists.* 2nd ed. Oxford: Oxford University Press.

Nash, Jennifer. 2008. "Re-thinking Intersectionality." *Feminist Review* 89: 1–15.

———. 2009. "Un-disciplining Intersectionality (Review Essay)." *International Feminist Journal of Politics* 11 (4): 587–93.

Nassar-McMillan, Mary Wyer, Maria Oliver-Hoyo, and Jennifer Schneider. 2011. "New Tools for Examining Undergraduate Students' STEM Stereotypes: Implications for Women and Other Underrepresented Groups." *New Directions for Institutional Research* 2011 (152): 87–98.

National Academy of Sciences. 2006. *Beyond Bias and Barriers: Fulfilling the Potential of Women in Academic Science and Engineering.* Washington, DC: National Academy Press.

National Science Foundation. 2002. "Demographics of the S&E Workforce." In *Science and Engineering Indicators 2002.* Arlington, VA.

———. 2011. "Demographics of the S&E Workforce." In *Science and Engineering Indicators 2011.* Arlington, VA.

———. 2012. "Demographics of the S&E Workforce." In *Science and Engineering Indicators 2012.* Arlington, VA.

Nelkin, Dorothy, and Susan Lindee. 1995. *The DNA Mystique: The Gene as Cultural Icon.* New York: W. H. Freeman.

Nelson, Donna. 2007. "An Analysis of Minorities in Science and Engineering Faculties in Research Universities," http://faculty-staff.ou.edu/N/Donna.J.Nelson-1/diversity/ Faculty_Tables_FY07/07Report.pdf.

Nerad, Maresi. 1999. The Academic Kitchen: A Social History of Gender Stratification at the University of California's Berkeley. Albany: State University of New York Press.

New York Times, Editorial. 2012. "Dream Act for New York," March 12.

Neyhfakh, Leon. 2011. "The Invasive Species War," Boston Globe, July 31.

Noble, David F. 1992. A World Without Women: The Christian Clerical Culture of Western Science. Oxford: Oxford University Press.

Norton, Bernard J. 1978. "Karl Pearson and Statistics: The Social Origins of Scientific Innovation." Social Studies of Science 8 (1): 3–34.

———. 1983. "Fisher's Entrance in Evolutionary Science: The Role of Eugenics." In Dimensions of Darwinism, edited by Marjorie Grene, 19–29. Cambridge: Cambridge University Press.

Nosek, B. A., F. L. Smyth, N. Sriram, N. M. Lindner, T. Devos, A. Ayala, et al. 2009. "National Differences in Gender-Science Stereotypes Predict National Sex Differences in Science and Math Achievement." Proceedings of the National Academy of Sciences 106: 10593–97.

Novas, Carlos, and Nikolas Rose. 2000. "Genetic Risk and the Birth of the Somatic Individual." Economy and Society 29 (4): 484–513.

Nye, Robert A. 1997. "Medicine and Science as Masculine 'Fields of Honor.'" Osiris 12: 60–79.

O'Brien, Dennis. 2004. "A Crusade to Stop the Voracious Fish," Baltimore Sun, April 30.

O'Brien, W. 2006. "Exotic Invasions, Nativism and Ecological Restoration: On the Persistence of a Contentious Debate." Ethics, Place and Environment 9: 63–77.

Odum, E. P. 1997. Ecology: A Bridge between Science and Society. Sunderland, MA: Sinauer Associates.

Ogilvie, Marilyn B., and Joy D. Harvey, eds. 2000. The Biographical Dictionary of Women in Science: Pioneering Lives from Ancient Times to the Mid-20th Century. New York: Routledge.

Oliver, L. R., R. E. Frans, and R. E. Talbert. 1976. "Field Competition between Tall Morningglory and Soybean. I. Growth Analysis." Weed Science 24 (5): 482–88.

Olwig, R. 2003. "Natives and Aliens in National Landscape." Landscape Research 28 (1): 61–74.

Omi, Michael, and Howard Winant. 1994. "Racial Formation." In Racial Formation in the United States: From the 1960's to 1990's. New York: Routledge.

Ong, Aihwa. 1999. Flexible Citizenship: The Cultural Logics of Transnationality. Durham, NC: Duke University Press.

———. 2006. "Higher Learning in Global Space." In Neoliberalism as Exception: Mutations in Citizenship and Sovereignty, 139–56. Durham, NC: Duke University Press.

Ong, Maria, Carol Wright, Lorelle L. Espinosa, and Gary Orfield. 2011, June. "Inside the Double Bind: A Synthesis of Empirical Research on Undergraduate and Graduate Women of Color in Science, Technology, Engineering, and Mathematics." Harvard Educational Review 6: 172–209.

Onishi, Norimitsu. 2012. "Crayfish to Eat, and to Clean the Water," *The New York Times*, July 12.

Orben, Robert. 2006. Quoted in Robert Byrne, *The 2,548 Best Things Anybody Ever Said*. New York: Simon & Schuster.

Paretti, Jonah. 1998. "Nativism and Nature: Rethinking Biological Invasions." *Environmental Values* 7 (2): 183–92.

Paul, Diane B. 1987. "'Our Load of Mutations' Revisited." *Journal of the History of Biology* 20 (3): 321–35.

———. 1995. *Controlling Human Heredity: 1865 to the Present*. Atlantic Highlands, NJ: Humanities Press.

———. 1998. "Did Eugenics Rest on an Elementary Mistake?" In *The Politics of Heredity: Essays on Eugenics, Biomedicine and the Nature-Nurture Debate*, 117–32. Albany: State University of New York Press.

———. 2001. "History of Eugenics." In *International Encyclopedia of Social and Behavioral Sciences*, edited by Neil Smelser and Paul Baltes, 4896–4901. Amsterdam: Elsevier.

———. 2007. "On Drawing Lessons from the History of Eugenics." In *Reprogenetics: Law, Policy, and Ethical Issues*, edited by Lori P. Knowles and Gregory E. Kaebnick. Baltimore: Johns Hopkins University Press.

Paul, Diane, and H. G. Spencer. 1995. "The Hidden Science of Eugenics." *Nature* 374 (6529): 302–4.

Pauly, Philip. 1993. "The Eugenics Industry: Growth or Restructuring?" *Journal of the History of Biology* 26 (1): 131–45.

———. 1996. "The Beauty and Menace of the Japanese Cherry Trees: Conflicting Visions of American Ecological Independence." *Isis* 87 (1): 51–73.

———. 2002, July. "Fighting the Hessian Fly: American and British Responses to Insect Invasion: 1776–1789." *Environmental History* 7 (3): 485–507.

PBS Newshour. 2012. "An Old Fashioned Strategy to Keep Asian Carp at Bay in the Great Lakes: Eat Them," July 26, http://www.pbs.org/newshour/bb/science/july-dec12/carp_07–26.html.

Philip, Kavita. 2004. *Civilizing Natures: Race, Resources, and Modernity in Colonial South India*. New Brunswick, NJ: Rutgers University Press.

Pigliucci, Massimo. 2012. "The One Paradigm To Rule Them All: Scientism and the Big Bang Theory." In *The Big Bang Theory and Philosophy: Rock, Paper, Scissors, Aristotle, Locke*. Oxford: Blackwell.

Pinker, Steven. 2002. *The Blank Slate: The Modern Denial of Human Nature*. New York: Penguin.

Pollan, Michael. 1994. "Against Nativism," *The New York Times Magazine*, May 15.

Porter, Theodore. 2005. *Karl Pearson: The Scientific Life in a Statistical Age*. Princeton, NJ: Princeton University Press.

Prakash, Gyan. 1999. *Another Reason: Science and the Imagination of Modern India*. Princeton, NJ: Princeton University Press.

Preston, Christopher D. 2009. "The Terms 'Native' and 'Alien'—A Bio-geographical Perspective." *Progress in Human Geography* 33: 702–13.

Pringle, Anne, James Bever, Monique Gardes, Jeri Parrent, Matthias Rillig, and John Klironomos. 2009. "Mycorrhizal Symbioses and Plant Invasions." *Annual Review of Ecology, Evolution, and Systematics* 40: 699–715.

Provine, W. 1971. *The Origins of Theoretical Population Genetics.* Chicago: University of Chicago Press.

———. 1988. *Sewall Wright and Evolutionary Biology.* Chicago: University of Chicago Press.

Puar, Jasbir. 2007. *Terrorist Assemblages: Homonationalism in Queer Times.* Durham, NC: Duke University Press.

———. 2012. "'I Would Rather Be a Cyborg Than a Goddess': Becoming-Intersectional of Assemblage Theory." *PhiloSOPHIA: A Journal of Feminist Philosophy* 2: 49–66.

Pycior, Helena. 1997. "Marie Curie: Time Only for Science and Family." In *A Devotion to Their Science: Pioneer Women of Radioactivity*, edited by Marelene F. Rayner-Canham and Geoffrey W. Rayner-Canham, 31–50. Philadelphia: Chemical Heritage Foundation.

Pycior, Helena, Nancy Slack, and Pnina Abir-Am, eds. 1996. *Creative Couples in the Sciences.* New Brunswick, NJ: Rutgers University Press.

Pysek, P., D. M. Richardson, J. Pergl, V. Jaros, Z. Sixtova, and E. Weber. 2008. "Geographical and Taxonomic Biases in Invasion Ecology." *Trends in Ecology and Evolution* 23: 237–44.

Pysek, P., D. M. Richardson, M. Rejmanek, G. L. Webster, M. Williamson, and J. Kirschener. 2004. "Alien Plants in Checklists and Floras: Towards Better Communication between Taxonomists and Ecologists." *Taxon* 53: 131–43.

Quinn, Susan. 1995. *Marie Curie: A Life.* New York: Simon & Schuster.

Rabinow, Paul, and Nikoas Rose. 2006. "Biopower Today." *BioSocieties* 1: 195–217.

Rader, Karen. 2004. *Making Mice: Standardizing Animals for American Biomedical Research, 1900–1955.* Princeton, NJ: Princeton University Press.

Raffles, Hugh. 2011. "Mother Nature's Melting Pot," *The New York Times*, April 2.

Raina, Dhruv, and S. Irfan Habib. 2004. *Domesticating Modern Science: A Social History of Science and Culture in Colonial India.* New Delhi: Tulika.

Rampell, Catherine. 2011. "Many with New College Degrees Find the Job Market Humbling," *New York Times*, May 18.

Ramsden, Edmund. 2002. "Carving Up Population Science: Eugenics, Demography and the Controversy over the 'Biological Law' of Population Growth." *Social Studies of Science* 32 (5/6): 857–99.

———. 2006. "Confronting the Stigma of Perfection: Genetic Demography, Diversity and the Quest for a Democratic Eugenics in the Post-war United States." Working Paper No. 12/06. On the Nature of Evidence: How Well Do "Facts" Travel? http://eprints.lse.ac.uk/22536/.

———. 2008. "Eugenics from the New Deal to the Great Society: Genetics, Demography and Population Quality." *Studies in the History and Philosophy of Biology and Biomedical Sciences* 39: 391–406.

———. 2009. "Confronting the Stigma of Eugenics: Genetics, Demography and the Problems of Population." *Social Studies of Science* 39 (6): 853–84.

Ramusack, Barbara. 1989. "Embattled Advocates: The Debate over Birth Control in India, 1920–1940." *Journal of Women's History* 1 (2): 34–64.

Raver, Anne. 2012. "Finding Flavor in the Weeds," *The New York Times,* July 6.

Reardon, Jenny. 2001. "The Human Genome Diversity Project: A Case Study in Coproduction." *Social Studies of Science* 31 (3): 357–88.

———. 2005. *Race to the Finish: Identity and Governance in an Age of Genomics.* Princeton, NJ: Princeton University Press.

———. 2011. "The Democratic, Anti-Racist Genome? Technoscience and the Limits of Liberalism." *Science as Culture* 21 (1).

Renn, Jürgen, and Robert Schulmann. 2000. *Albert Einstein, Mileva Maric: The Love Letters.* Princeton, NJ: Princeton University Press.

Resnick, Michael. 1997. *Turtles, Termites and Traffic Jams: Explorations in Massively Parallel Microworlds.* Cambridge, MA: MIT Press.

Reuters. 2002. "Remains of Abused South African Woman Given Final Resting Place," *The New York Times,* August 9, http://www.nytimes.com/2002/08/09/international/09RTRS-VENU.html.

Rich, Benjamin. 2012. "The Gated Community Mentality," *The New York Times,* March 29.

Richardson, D. M. 2011. "Invasion Science: The Roads Travelled and the Roads Ahead." In *Fifty Years of Invasion Ecology: The Legacy of Charles Elton,* edited by David M. Richardson, 397–407. Oxford: Blackwell.

Richardson, D. M., P. M. Holmes, K. J. Esler, et al. 2007. "Riparian Zones—Degradation, Alien Plant Invasions and Restoration Prospects." *Diversity and Distributions* 13: 126–39.

Richardson, D. M., and P. Pysek. 2008. "Invasion Ecology and Restoration Ecology: Parallel Evolution in Two Fields of Endeavor." In *Fifty Years of Invasion Ecology: The Legacy of Charles Elton,* edited by David M. Richardson, 61–70. Oxford: Blackwell.

Richardson, Sarah. 2012. "Sexing the X: How the X Became the 'Female Chromosome.'" *Signs* 37 (4): 909–33.

Ridley, Mathew. 2000. *Genome: The Autobiography of a Species in 23 Chapters.* New York: Harper Perennial.

Ringle, Ken. 2002. "Stop That Fish!" *Washington Post,* July 3, C01.

Roberts, Dorothy. 1997. *Killing the Black Body: Race, Reproduction, and the Meaning of Liberty.* New York: Pantheon.

———. 2008. "Is Race Based Medicine Good for Us? African American Approaches to Race, Biomedicine, and Equality." *Journal of Law, Medicine and Ethics* 36 (3): 537–45.

———. 2009. "Race, Gender, and Genetic Technologies: A New Reproductive Dystopia?" *Signs* 34 (4): 783–803.

———. 2012. *Fetal Invention: How Science, Politics, and Big Business Re-create Race in the Twenty-first Century.* New York: The New Press.

Roberts, Philip D., Hilda Diaz-Soltero, David J. Hemming, Martin J. Parr, Nicola H. Wakefield, and Holly J. Wright. 2013. "What Is the Evidence That Invasive Species Are a

Significant Contributor to the Decline or Loss of Threatened Species? A Systematic Review Map." *Environmental Evidence*, 2 (5), http://www.environmentalevidencejournal.org/content/2/1/5.

Robichaux, Mark. 2000. "Alien Invasion: Plague of Asian Eels Highlights Damage from Foreign Species," *Wall Street Journal*, September 27, 12A.

Robinson, Carol. 2011. "Women in Science: In Pursuit of Female Chemists." *Nature* 476: 273–75.

Rodger, D., J. Stokes, and J. Ogilvie. 2003. *Heritage Trees of Scotland*. London: The Tree Council.

Rodman, J. 1993: "Restoring Nature: Natives and Exotics." In *In the Nature of Things: Language, Politics and the Environment*, edited by J. Bennett and W. Chaloupka, 139–53. Minneapolis: University of Minnesota Press.

Rolin, Kristina. 2004, Winter. "Three Decades of Feminism in Science: From 'Liberal Feminism' and 'Difference Feminism' to Gender Analysis of Science." *Hypatia* 19 (1): 292–96.

Roll-Hansen, Nils. 2010. "Eugenics and the Science of Genetics." In *The Oxford Handbook of the History of Eugenics*, edited by Alison Bashford and Philippa Levine, 80–97. Oxford: Oxford University Press:.

Rone, Adam. 2008, July. "Nature Wars, Culture Wars: Immigration and Environmental Reform in the Progressive Era." *Environmental History* 13 (3): 432–53.

Roosth, Sophia, and Astrid Schrader, eds. 2012. "Feminist Theory Out of Science." Special issue, *differences* 23 (3).

Rose, Hilary. 1994. *Love, Power and Knowledge: Towards a Feminist Transformation of the Sciences*. Cambridge: Polity.

———. 1997. "Good-bye Truth, Hello Trust: Prospects for Feminist Science and Technology Studies at the Millennium?" In *Science and the Construction of Women*, edited by Mary Maynard. New York: Routledge.

———. 2007. "Eugenics and Genetics: The Conjoint Twins? New Formations," March 22, http://business.highbeam.com/150795/article-1G1–162578234/eugenics-and-genetics-conjoint-twins.

Rose, Nikolas. 2006. *The Politics of Life Itself: Biomedicine, Power, and Subjectivity in the Twenty-First Century*. Princeton, NJ: Princeton University Press.

Rosser, Sue V. 1990. *Female-Friendly Science: Applying Women's Studies Methods and Theories to Attract Students*. New York: Pergamon.

———, ed. 1995. *Teaching the Majority: Breaking the Gender Barrier in Science, Mathematics, and Engineering*. New York: Teachers College Press.

———. 1997. *Re-engineering Female Friendly Science*. New York: Teachers College Press.

———. 2004a. *The Science Glass Ceiling: Academic Women Scientists and the Struggle to Succeed*. New York: Routledge.

———, ed. 2004b. "Using POWRE to ADVANCE: Institutional Barriers Identified by Women Scientists and Engineers." *NWSA Journal* 13 (1): 139–49.

———, ed. 2008. *Women, Science, and Myth: Gender Beliefs from Antiquity to the Present*. Santa Barbara, CA: ABC-CLIO.

Rosser, Sue V., and Mark Zachary Taylor. 2009. "Why Are We Still Worried about Women in Science?" *Academe* 95 (3): 7–10.

Rossiter, Margaret W. 1982. *Women Scientists in America: Struggles and Strategies to 1940.* Baltimore: Johns Hopkins University Press.

———. 1993. "The Matilda Effect in Science." *Social Studies of Science* 23: 325–41.

———. 1995. *Women Scientists in America: Before Affirmative Action, 1940–1972.* Baltimore: Johns Hopkins University Press.

Roy, Deboleena. 2004. "Feminist Theory in Science: Working Toward a Practical Transformation." *Hypatia* 19 (1): 255–79.

———. 2008. "Asking Different Questions: Feminist Practices for the Natural Sciences." *Hypatia* 23 (4): 134–57.

Rubin, David. 2012. "'An Unnamed Blank That Craved a Name': A Genealogy of Intersex as Gender." *Signs* 37 (4): 883–908.

Ruse, Michael. 2005. "The Darwinian Revolution, as Seen in 1979 and as Seen Twenty-Five Years Later in 2004." *Journal of the History of Biology* 38 (1): 3–17.

Rushdie, Salman. 1988. *The Satanic Verses.* London: Viking.

Sagoff, Mark. 2000, June. "Why Exotic Species Are Not as Bad as We Fear." *Chronicle of Higher Education* 46 (42): B7.

———. 2005. "Do Non-native Species Threaten the Natural Environment?" *Journal of Agricultural Environmental Ethics* 18: 215–36.

———. 2006. "Environmental Ethics and Environmental Science." In *Environmental Ethics and International Policy*, edited by Henk ten Have, 145–61. Paris: UNESCO.

Sahlins, Marshall. 2000. "'Sentimental Pessimism' and Ethnographic Experience: Or, Why Culture Is Not a Disappearing Object." In *Biographies of Scientific Objects*, edited by Lorraine Daston, 158–202. Chicago: University of Press.

Said, Edward. 1979. *Orientalism.* New York: Vintage.

Samerski, Silja. 2009. "Genetic Counseling and the Fiction of Choice: Taught Self-Determination as a New Technique of Social Engineering." *Signs* 32 (4): 735–61.

Sandler, Bernice R. 1986. "The Campus Climate Revisited: Chilly for Women Faculty, Administrators, and Graduate Students." In *Project on the Status and Education of Women*, edited by Bernice R. Sandler. Washington, DC: Association of American Colleges.

Sax, Linda J. 2001, Winter. "Undergraduate Science Majors: Gender Differences in Who Goes to Graduate School." *The Review of Higher Education* 24 (2): 153–72.

Sayre, Anne. 1975. *Rosalind Franklin and DNA.* New York: Norton.

Sayres, Janet. 1982. *Biological Politics: Feminist and Anti-feminist Perspectives.* London: Tavistock.

Scantlebury, Kate, ed. 2010. *Revisioning Science Education from Feminist Perspectives: Challenges, Choices, and Careers.* New York: Sense.

Scher, Lauren, and Fran O'Reilly. 2009. "Professional Development for K–12 Math and Science Teachers: What Do We Really Know?" *Journal of Research on Educational Effectiveness* 2 (3): 209–49.

Schiebinger, Londa. 1989. *The Mind Has No Sex? Women in the Origins of Modern Science.* Cambridge, MA: Harvard University Press.

———. 1993. *Nature's Body: Gender in the Making of Modern Science.* Boston: Beacon.

———. 1999. *Has Feminism Changed Science?* Cambridge, MA: Harvard University Press.

———. 2003. "Introduction: Feminism Inside the Sciences." *Signs* 28 (3): 859–66.

———. 2004. "Feminist History of Colonial Science." *Hypatia* 19 (1): 233–53.

———, ed. 2008. *Gendered Innovations in Science and Engineering.* Stanford, CA: Stanford University Press.

Schweber, S. S. 1978. "The Genesis of Natural Selection—1838: Some Further Insights." *BioScience* 28 (5): 321–26.

Scott, Joan. 1991. "The Evidence of Experience." *Critical Inquiry* 17 (4): 773–97.

Scott, Timothy Lee, and Stephen Buhmer. 2010. "Invasive Plant Medicine: The Ecological Benefits and Healing Abilities of Invasives." Rochester, VT: Inner Traditions.

Searle, G. R. 1976. *Eugenics and Politics in Britain, 1900–1914.* Leiden, Neth.: Noordhof International.

Selden, Steven. 2005. "Transforming Better Babies into Fitter Families: Archival Resources and the History of the American Eugenics Movement, 1908–1930." *Proceedings of the American Philosophical Society* 149 (2): 199–225.

Settles, Isis H., Lilia M. Cortina, Janet Malley, and Abigail J. Stewart. 2006. "The Climate for Women in Academic Science: The Good, the Bad, and the Changeable." *Psychology of Women Quarterly* 30: 47–58.

Sharpley-Whiting, Denean T. 1996. "The Dawning of Racial-Sexual Science: A One Woman Showing, a One Man Telling." *Ethnography in French Literature* 33: 115–28.

Shteir, Ann. B. 1999. *Cultivating Women, Cultivating Science: Flora's Daughters and Botany in England, 1760–1860.* Baltimore: Johns Hopkins University Press.

Simberloff, Daniel. 2003. "Confronting Invasive Species: A Form of Xenophobia?" *Biological Invasions* 5: 179–92.

Simberloff, Daniel, et al. 2011. "Non-natives: 141 Scientists Object." *Nature* 475 (36). DOI: 10.1038/457036a

Sime, Ruth Lewin. 1997. *Lise Meitner: A Life in Physics.* Berkeley: University of California Press.

Sivaramakrishnan, K. 2011. "Thin Nationalism: Nature and Public Intellectualism in India." *Contributions to Indian Sociology* 45: 85–111.

Slaughter, Sheila, and Larry L. Leslie. 2007. "Expanding and Elaborating the Concept of Academic Capitalism." *Organization* 8 (2): 154–61.

Slaughter, Sheila, and Gary Rhoades. 2000, Spring/Summer. "The New Liberal University." *New Labor Forum* 6: 73–79.

———. 2004. *Academic Capitalism and the New Economy: Markets, State and Higher Education.* Baltimore: Johns Hopkins University Press.

Slobodkin, L. B. 2001. "The Good, the Bad and the Reified." *Evolutionary Ecology Research* 3:1–13.

Smocovitis, V. B. 1992. "Unifying Biology: The Evolutionary Synthesis and Evolutionary Biology." *Journal of the History of Biology* 25: 58–59.

Smout, Chris. T. 2003. "The Alien Species in 20th Century Britain: Constructing a New Vermin." *Landscape Research* 29 (1): 11–20.

Sober, Elliott. 1980. "Evolution, Population Thinking, and Essentialism." *Philosophy of Science* 47 (3): 350–83.

Somerville, Siobhan. 2005. "Notes Toward a Queer History of Naturalization." *American Quarterly* 57 (3): 659–75.

Sonnert, Gerhard, and Mary Frank Fox. 2012, January/February. "Women, Men, and Academic Performance in Science and Engineering: The Gender Difference in Undergraduate Grade Point Averages." *Journal of Higher Education* 83 (1): 73–101.

Sonnert, Gerhard, and Gerald J. Holton. 1995. *Who Succeeds in Science? The Gender Dimension.* New Brunswick, NJ: Rutgers University Press.

———. 1996. "Career Patterns of Men and Women in the Sciences." *American Scientist* 84: 63–71.

Soulé, M. E. 1990. "The Onslaught of Alien Species, and Other Challenges in the Coming Decades." *Conservation Biology* 4: 233–39.

Soulé, M. E., and G. Lease. 1995. *Reinventing Nature: Responses to Postmodern Deconstruction.* Washington, DC: Island.

Spelke, Elizabeth S. 2005. "Sex Differences in Intrinsic Aptitude for Mathematics and Science?: A Critical Review." *American Psychology* 60: 950–58.

Srivastava, R. C. 1983. "A Taxonomic Study of the Genus *Ipomoea L.* (*Convolvulaceae*) in Madhya Pradesh." *Journal of Economic and Taxonomic Botany* 4: 765–75.

Stampe, Elizabeth, and Curtis Daehler. 2003. "Mycorrhizal Species Identity Affects Plant Community Structure and Invasion: A Microcosm Study." *Oikos* 100: 362–72.

Steinem, Gloria. 2005. Acceptance speech cited in Taylor, http://articles.chicagotribune.com/2005-05-18/features/0505180315_1_minority-owned-firms-women-s-movement-anger.

Stengers, Isabelle. 2000. *The Invention of Modern Science.* Minneapolis: University of Minnesota Press.

———. 2011. *Thinking with Whitehead: A Free and Wild Creation of Concepts.* Cambridge, MA: Harvard University Press.

Stepan, Nancy. 1982. *The Idea of Race in Science: Great Britain, 1800–1960.* Hamden, CT: Anchor.

———. 1985. "Biological Degeneration: Races and Proper Places." In *Degeneration: The Dark Side of Progress,* edited by Edward Chamberlin and Sander L. Gilman. New York: Columbia University Press.

———. 1986. "Race and Gender: The Role of Analogy in Science." *Isis* 77 (2): 261–77.

Stern, Alexandra Minna. 2005. *Eugenic Nation: Faults and Frontiers of Better Breeding in Modern America.* Berkeley: University of California Press.

———. 2010. "Gender and Sexuality: A Global Tour and Compass." In *The Oxford Handbook of the History of Eugenics,* edited by Alison Bashford and Philippa Levine, 173–91. Oxford: Oxford University Press.

Stewart, Abigail J., Janet E. Malley, and Danielle Lavaque-Manty, eds. 2007. *Transforming Science and Engineering: Advancing Academic Women.* Ann Arbor: University of Michigan Press.

Stewart, Barbara. 2001. "The Invasion of the Woodland Soil Snatchers," *The New York Times,* April 24, 1B.

Stoler, Ann, C. McGranahan, and P. C. Perdue. 2008. *Imperial Formations.* Santa Fe, NM: School for Advanced Research Press.

Subrahmanyan, Lalita. 1998. *Women Scientists in the Third World: The Indian Experience.* New Delhi: Sage.

Subramaniam, Banu. 1994. "Maintenance of the Flower Color Polymorphism at the W Locus in the Common Morning Glory, *Ipomoea purpurea.*" Ph.D. diss., Duke University.

———. 1998. "A Contradiction in Terms: Life as a Feminist Scientist." *Women's Review of Books* 15 (5): 25–26.

———. 2001. "The Aliens Have Landed! Reflections on the Rhetoric of Biological Invasions." *Meridians: feminism, race, transnationalism* 2 (1): 26–40.

———. 2009. "Moored Metamorphoses: A Retrospective Essay on Feminist Science Studies." *Signs* 34 (4): 951–80.

Subramaniam, Banu, Rebecca Dunn, and Lynn Broaddus. 1992. "'Sir'vey or 'Her'vey." In *Engaging Feminism: Students Speak Up and Speak Out,* edited by Jean O'Barr and Mary Wyer. Chicago: University of Chicago Press.

Subramaniam, Banu, and Mark Rausher. 2000. "Balancing Selection on a Floral Polymorphism." *Evolution* 54 (2): 691–95.

Subramaniam, Banu, and Mary Wyer. 1998. "Assimilating the 'Culture of No Culture' in Science: Feminist Interventions in (De)Mentoring Graduate Women." *Feminist Teacher* 12 (1): 12–28.

Sur, Abha. 2001. "Dispersed Radiance: Women Scientists in C. V. Raman's Laboratory." *Meridians: Feminism, Race, Transnationalism* 1 (2): 95–127.

———. 2011. *Dispersed Radiance: Caste, Gender, and Modern Science in India.* New Delhi: Navanyana.

Swift, Katherine. 2008, May. "Sinister Science: Eugenics, Nazism, and the Technocratic Rhetoric of the Human Betterment Foundation." *Lore* 6 (2): 1–11.

Takeshita, Chikako. 2011 *The Global Biopolitics of the IUD: How Science Constructs Contraceptive Users and Women's Bodies.* Cambridge, MA: MIT Press.

Taylor, Mark. C. 2010. *Crisis on Campus: A Bold Plan for Reforming Our Colleges and Universities.* New York: Knopf.

———. 2011, April. "Reform the PhD System or Close It Down." *Nature* 472: 261.

Teresi, Dick. 2001. *Lost Discoveries: Ancient Roots of Modern Science—From the Babylonians to the Mayans.* New York: Simon & Schuster.

Terry, Jennifer. 1999. *An American Obsession: Science, Medicine, and Homosexuality in Modern Society.* Chicago: University of Chicago Press.

Theodoropoulos, D. I. 2003. *Invasion Biology: Critique of a Pseudoscience.* Blythe, CA: Avvar.

Todd, Kim. 2001. *Tinkering with Eden: A Natural History of Exotics in America*. New York: Norton.

Tomes, Nancy. 2000, February. "The Making of a Germ Panic, Then and Now." *American Journal of Public Health* 90 (2): 191–99.

Townsend, M. 2005. "Is the Social Construction of Native Species a Threat to Biodiversity?" *ECOS* 26: 1–9.

Traweek, Sharon. 1992. *Beamtimes and Lifetimes: The World of High Energy Physicists*. Cambridge, MA: Harvard University Press.

———. 1993. "Cultural Differences in High-Energy Physics: Contrasts between Japan and the United States." In *The "Racial" Economy of Science: Toward a Democratic Future*, edited by Sandra Harding, 398–407. Bloomington: Indiana University Press.

Tredoux, Gavan. 2001. *Two Geneticists: J.B.S. Haldane and C.D. Darlington*, http://www.mail-archive.com/ctrl@listserv.aol.com/msg73477.html.

Tsing, Anna Lowenhaupt. 1994. "Empowering Nature, or Some Gleanings in Bee Culture." In *Naturalizing Power: Essays in Feminist Cultural Analysis*, edited by Sylvia Yanagisako and Carol Delaney, 113–43. New York: Routledge.

Tuana, Nancy, ed. 1989. *Feminism and Science*. Bloomington: Indiana University Press.

Turda, Marius. 2010. *Modernism and Eugenics*. Basingstoke, UK: Palgrave Macmillan.

United Press International. 1998. "Native Species Invaded." *ABC News*, March 16.

U.S. Department of Agriculture. 2012. "Natives, Invasive, and Other Plant-Related Definitions," http://www.ct.nrcs.usda.gov/plant_definitions. html.

———. n.d. "Plants." http://www.invasivespeciesinfo.gov/plants/main.shtml.

U.S. Geological Survey Midcontinent Ecological Science Center. 1999, May 13. "USGS Research Upsets Conventional Wisdom on Invasive Species Invasions." News release.

Valent, Barbara. 1990. "Rice Blast as a Model System for Plant Pathology." *Phytopathology* 80 (1): 33–36.

Valentine, G. 2004. "Geography and Ethics: Questions of Considerability and Activism in Environmental Ethics." *Progress in Human Geography* 28: 258–64.

Valian, Virginia. 1999. *Why So Slow? The Advancement of Women*. Cambridge, MA: MIT Press.

Valla, Jeffrey M., and Wendy M. Williams. 2012. "Increasing Achievement and Higher-Education Representation of Under-Represented Groups in Science, Technology, Engineering, and Mathematics Fields: A Review of Current K–12 Intervention Programs." *Journal of Women and Minorities in Science* 18 (1): 21–53.

Van Driesche, Jason, and Roy Van Driesche. 2000. *Nature Out of Place: Biological Invasions in the Global Age*. Washington, DC: Island.

Veit, Helen Zoe. 2011. "Time to Revive Home Ec," *The New York Times*, September 5.

Venter, O., N. N. Brodeur, L. Nemiroff, B. Belland, I. J. Dolinsek, and J. W. A. Grant. 2006. "Threats to Endangered Species in Canada." *Bioscience* 56: 903–10.

Vermeij, G. J. 2005. "Invasion as Expectation: A Historical Fact of Life." In *Species Invasions: Insights into Ecology, Evolution, and Biogeography*, edited by D. F. Sax, J. J. Stachowicz, and S. D. Gaines, 315–36. Sunderland, MA: Sinauer Associates.

Verran, Helen. 2001. *Science and an African Logic*. Chicago: University of Chicago Press.

Verrengia, Joseph B. 1999. "When Ecologists Become Killers." *MSNBC News*, October 4, www.msnbc.com/news.

———. 2000. "Some Species Aren't Welcome." *ABC News*, September 27, www.abcnews .go.com.

Veuille, Michel. 2010. "Darwin and Sexual Selection: 100 Years of Misunderstanding." *C. R. Biologies* 333 (2): 145–56.

Vining, D. R. 1983. "Fertility Differentials and the Status of Nations: A Speculative Essay on Japan and the West." In *Intelligence and National Achievement,* edited by R. B. Cattell. Washington, DC: Cliveden.

Viswanathan, Shiv. 2002. "The Laboratory and the World: Conversations with C. V. Seshadri," *Economic and Political Weekly,* June 1, 2163–70.

Viswesaran, Kamala. 1994. *Fictions of Feminist Ethnography.* Minneapolis: University of Minnesota Press.

Waggoner, Miranda. R. 2012. "Motherhood Preconceived: The Emergence of the Preconception Health and Health Care Initiative." *Journal of Health Politics, Policy and Law* 38 (2): 345–71.

Wajcman, Judy. 1991. *Women Confront Technology.* University Park: Penn State University Press.

Walsh, Denis. 2006. "Evolutionary Essentialism." *British Journal of Philosophy of Science* 57 (2): 425–48.

Walters, Suzanna. 1996. "From Here to Queer." *Signs* 21 (4): 857.

Wanzo, Rebecca. 2009. "In the Shadows of Anarcha." In *This Suffering Will Not Be Televised.* New York: State University of New York Press.

Warren, Charles R. 2007. "Perspectives on the 'Alien' versus 'Native' Species Debate: A Critique of Concepts, Language and Practice." *Progress in Human Geography* 31 (4): 427–46.

Wayne, Marta L. 2000. "Walking a Tightrope: The Feminist Life of a *Drosophila* Biologist." *NWSA Journal* 12 (3): 139–50.

Weiner, H. 1996. "Congress Threatens Wild Immigrants." *Earth Island Journal* 11 (4): 19.

Weinstein, Naomi. 1977. "'How Can a Little Girl Like You Teach a Great Big Class of Men?' the Chairman Said, and Other Adventures of a Woman in Science." In *Working it Out: 23 Women Writers, Artists, Scientists, and Scholars Talk About Their Lives and Work,* edited by Sara Ruddick and Pamela Daniels. New York: Pantheon.

Wilson, Elizabeth A. 1998. *Neural Geographies: Feminism and the Microstructure of Cognition.* New York: Routledge.

———. 2004. *Psychosomatic: Feminism and the Neurological Body.* Durham, NC: Duke University Press.

Winsor, M. P. 2003. "Non-essentialist Methods in Pre-Darwinian Taxonomy." *Biology and Philosophy* 18: 387–400.

Woodhouse. Edward I. 2005. "Reconstructing Technological Society by Taking Social Construction Even More Seriously." *Social Epistemology* 19: 2–3.

Woods, Mark, and Paul Veatch Moriarty. 2001. "Strangers in a Strange Land: The Problem of Exotic Species." *Environmental Values* 10 (2): 163–91.

Wong, Tama Matsuoka, and Eddy Leroux. 2012. *Foraged Flavor: Finding Fabulous Ingredients in Your Backyard or Farmer's Market*. New York: Clarkson Potter.

Wright, S. F., and A. Upadhyaya. 1998. "A Survey of Soils for Aggregate Stability and Glomalin, a Glycoprotein Produced by Hyphae of Arbuscular Mycorrhizal Fungi." *Plant and Soil* 198: 97–107.

Wyer, Mary. 1993. *NSF Model Project, Women in Science and Engineering*. Durham, NC: Duke University Press.

Wyer, Mary, Mary Barbercheck, Donna Giesman, Hatice Orun Ozturk, and Marta Wayne, eds. 2008. *Women, Science, and Technology: A Reader in Feminist Science Studies*. 2nd ed. New York: Routledge.

Wylie, Alison. 2011. "What Knowers Know Well: Women, Work and the Academy." In *Feminist Epistemology and Philosophy of Science: Power in Knowledge*. New York: Springer.

Xie, Yu, and Kimberlee Shauman. 2003. *Women in Science: Career Progress and Outcomes*. Cambridge, MA: Harvard University Press.

Yeh, Ling-Ling. 1995. "U.S. Can't Handle Today's Tide of Immigrants." *Christian Science Monitor* 87 (81): 19.

Young, Robert M. 1990. "Darwinism and the Division of Labour." *Science as Culture* 9: 110–24.

Yuval-Davis, Nira. 2006. "Intersectionality and Feminist Politics." *European Journal of Women's Studies* 13 (3): 193–209.

Zachariah, Benjamin. 2001. "The Uses of Scientific Argument: The Case of 'Development' in India c. 1930–1950." *Economic and Political Weekly* 36 (39): 3695.

Zakaib, Gwyneth Dickey. 2011. "Science Gender Gap Probed." *Nature* 470: 153.

Zerner, Charles, ed. 2000. *People, Plants, and Justice: The Politics of Nature Conservation*. New York: Columbia University Press.

Zhang, Qian, Ruyi, Yang, Jianjun Tang, Haishui Yang, Shuijin Hu, and Xin Chen. 2010. "Positive Feedback between Mycorrhizal Fungi and Plants Influences Invasion Success and Resistance to Invasion." *PLos One* 5 (8): 1–10.

Zimmer, Oliver. 1998. "In Search of Natural Identity: Alpine Landscape and the Reconstruction of the Swiss Nation." *Comparative Studies in Society and History* 40 (4): 637–65.

Zuckerman, Harriet, and Jonathan R. Cole. 1991. *The Outer Circle: Women in the Scientific Community*. New York: Norton.

Zufall, R. A., and M. D. Rausher. 2003. "The Genetic Basis of Flower Color Polymorphism in the Common Morning Glory, *Ipomoea purpurea*." *Journal of Heredity* 94 (6): 442–48.

Index

Banu Subramaniam is an associate professor of women, gender, sexuality studies at the University of Massachussetts, Amherst, and a coeditor of *Feminist Studies: A New Generation and Making Threats: Biofears and Environmental Anxieties.*

The University of Illinois Press
is a founding member of the
Association of American University Presses.

Composed in 10.5/13 Arno Pro
with Avenir display
by Jim Proefrock
at the University of Illinois Press
Manufactured by Sheridan Books, Inc.

University of Illinois Press
1325 South Oak Street
Champaign, IL 61820-6903
www.press.uillinois.edu